Developing Affordable and Accessible Community-Based Housing for Vulnerable Adults

PROCEEDINGS OF A WORKSHOP

Joe Alper, Karen Anderson, and Sarah Domnitz, *Rapporteurs*

Forum on Aging, Disability, and Independence

Roundtable on the Promotion of Health Equity and the Elimination of Health Disparities

Board on Health Sciences Policy

Board on Population Health and Public Health Practice

Health and Medicine Division

Division of Behavioral and Social Sciences and Education

The National Academies of
SCIENCES • ENGINEERING • MEDICINE

THE NATIONAL ACADEMIES PRESS
Washington, DC
www.nap.edu

THE NATIONAL ACADEMIES PRESS   500 Fifth Street, NW   Washington, DC 20001

This activity was supported by AARP; AARP Foundation; Administration for Community Living (Contract No. HHSP233201400020B/HHSP23337051); Aetna Foundation; Alliance for Home Health Quality and Innovation; The American Geriatrics Society; The Colorado Trust; Consumer Technology Association Foundation; The Gerontological Society of America; Health Resources and Services Administration (Contract No. HHSH250201500001I/HHSH25034002T; HHSH250200976014I/HHSH25034010T); The John A. Hartford Foundation; Kaiser Permanente; The Kresge Foundation; LeadingAge; the MacArthur Foundation; Merck & Co., Inc.; Methodist Healthcare Ministries; National Institute on Disability, Independent Living, and Rehabilitation Research; PHI; Qualcomm Inc. (Contract No. NAT-301711); U.S. Department of Defense (Contract No. HU0001-15-1-0007); U.S. Department of Housing and Urban Development; and U.S. Department of Veterans Affairs (Contract No. VA119-16-P-0111/PO101-C60259; VA268-15-C-0057). Any opinions, findings, conclusions, or recommendations expressed in this publication do not necessarily reflect the views of any organization or agency that provided support for the project.

International Standard Book Number-13:   978-0-309-45980-8
International Standard Book Number-10:   0-309-45980-X
Digital Object Identifier: https://doi.org/10.17226/24787

Additional copies of this publication are available for sale from the National Academies Press, 500 Fifth Street, NW, Keck 360, Washington, DC 20001; (800) 624-6242 or (202) 334-3313; http://www.nap.edu.

Copyright 2017 by the National Academy of Sciences. All rights reserved.

Printed in the United States of America

Suggested citation: National Academies of Sciences, Engineering, and Medicine. 2017. *Developing affordable and accessible community-based housing for vulnerable adults: Proceedings of a workshop.* Washington, DC: The National Academies Press. https://doi.org/10.17226/24787.

The National Academies of
SCIENCES · ENGINEERING · MEDICINE

The **National Academy of Sciences** was established in 1863 by an Act of Congress, signed by President Lincoln, as a private, nongovernmental institution to advise the nation on issues related to science and technology. Members are elected by their peers for outstanding contributions to research. Dr. Marcia McNutt is president.

The **National Academy of Engineering** was established in 1964 under the charter of the National Academy of Sciences to bring the practices of engineering to advising the nation. Members are elected by their peers for extraordinary contributions to engineering. Dr. C. D. Mote, Jr., is president.

The **National Academy of Medicine** (formerly the Institute of Medicine) was established in 1970 under the charter of the National Academy of Sciences to advise the nation on medical and health issues. Members are elected by their peers for distinguished contributions to medicine and health. Dr. Victor J. Dzau is president.

The three Academies work together as the **National Academies of Sciences, Engineering, and Medicine** to provide independent, objective analysis and advice to the nation and conduct other activities to solve complex problems and inform public policy decisions. The National Academies also encourage education and research, recognize outstanding contributions to knowledge, and increase public understanding in matters of science, engineering, and medicine.

Learn more about the National Academies of Sciences, Engineering, and Medicine at **www.nationalacademies.org**.

The National Academies of
SCIENCES • ENGINEERING • MEDICINE

Consensus Study Reports published by the National Academies of Sciences, Engineering, and Medicine document the evidence-based consensus on the study's statement of task by an authoring committee of experts. Reports typically include findings, conclusions, and recommendations based on information gathered by the committee and the committee's deliberations. Each report has been subjected to a rigorous and independent peer-review process and it represents the position of the National Academies on the statement of task.

Proceedings published by the National Academies of Sciences, Engineering, and Medicine chronicle the presentations and discussions at a workshop, symposium, or other event convened by the National Academies. The statements and opinions contained in proceedings are those of the participants and are not endorsed by other participants, the planning committee, or the National Academies.

For information about other products and activities of the National Academies, please visit www.nationalacademies.org/about/whatwedo.

PLANNING COMMITTEE FOR A WORKSHOP ON AFFORDABLE AND ACCESSIBLE HOUSING FOR VULNERABLE OLDER ADULTS AND PEOPLE WITH DISABILITIES LIVING IN THE COMMUNITY[1]

TERESA L. LEE (*Chair*), Executive Director, Alliance for Home Health Quality and Innovation
PAULA CARDER, Assistant Professor, Portland State University
ELENA FAZIO, Social Science Analyst, Office of Performance and Evaluation, Administration for Community Living
OCTAVIO MARTINEZ, Executive Director, Hogg Foundation for Mental Health, and The University of Texas at Austin
ANAND PAREKH, Chief Medical Advisor, Bipartisan Policy Center
CRAIG RAVESLOOT, Research Professor of Psychology, and Director, Research and Training, Center on Disability in Rural Communities, University of Montana
EMILY ROSENOFF, Senior Policy Analyst, Office of the Assistant Secretary for Planning and Evaluation, U.S. Department of Health and Human Services
UCHENNA S. UCHENDU, Chief Officer, Office of Health Equity, U.S. Department of Veterans Affairs
GLEN W. WHITE, Professor, Applied Behavior Science, and Director, Research and Training, Center on Independent Living, University of Kansas

Project Staff

SARAH DOMNITZ, Director, Forum on Aging, Disability, and Independence
KAREN ANDERSON, Director, Roundtable on the Promotion of Health Equity and the Elimination of Health Disparities
HILARY BRAGG, Program Coordinator
ANNA MARTIN, Senior Program Assistant
ANDREW M. POPE, Director, Board on Health Sciences Policy
ROSE MARIE MARTINEZ, Director, Board on Population Health and Public Health Practice

Consultant

JOE ALPER, Consulting Writer

---

[1] The National Academies of Sciences, Engineering, and Medicine's planning committees are solely responsible for organizing the workshop, identifying topics, and choosing speakers. The responsibility for the published Proceedings of a Workshop rests with the workshop rapporteurs and the institution.

FORUM ON AGING, DISABILITY, AND INDEPENDENCE[1]

TERRY T. FULMER (*Co-Chair*), The John A. Hartford Foundation
FERNANDO TORRES-GIL (*Co-Chair*), Luskin School of Public Affairs, University of California, Los Angeles
JAMES C. APPLEBY, The Gerontological Society of America
KENNETH BRUMMEL-SMITH, The American Geriatrics Society, and Florida State University College of Medicine
MARGARET L. CAMPBELL, Campbell & Associates, Consultants in Aging, Disability, and Technology Research and Policy
THOMAS E. EDES, U.S. Department of Veterans Affairs
ROBERT ESPINOZA, Paraprofessional Healthcare Institute
STEVE EWELL, Consumer Technology Association Foundation
ELENA FAZIO, Administration for Community Living
ROBERT JARRIN, Qualcomm Inc.
TERESA L. LEE, Alliance for Home Health Quality and Innovation
KATIE SMITH SLOAN, LeadingAge
JACK W. SMITH, U.S. Department of Defense
ERWIN TAN, AARP
MICHELLE M. WASHKO, Health Resources and Services Administration

Forum on Aging, Disability, and Independence Staff

SARAH DOMNITZ, Director, Forum on Aging, Disability, and Independence
HILARY BRAGG, Program Coordinator
GOOLOO WUNDERLICH, Senior Program Officer
ANDREW M. POPE, Director, Board on Health Sciences Policy

---

[1] The National Academies of Sciences, Engineering, and Medicine's forums and roundtables do not issue, review, or approve individual documents. The responsibility for the published Proceedings of a Workshop rests with the workshop rapporteurs and the institution.

ROUNDTABLE ON THE PROMOTION OF HEALTH EQUITY AND THE ELIMINATION OF HEALTH DISPARITIES[1]

ANTONIA M. VILLARRUEL (*Chair*), University of Pennsylvania
PATRICIA BAKER, Connecticut Health Foundation
JULIE A. BALDWIN, Northern Arizona University
GILLIAN BARCLAY, Independent Consultant
REBECCA BRUNE, Methodist Healthcare Ministries of South Texas, Inc.
NED CALONGE, The Colorado Trust
LUTHER T. CLARK, Merck & Co., Inc.
FRANCISCO GARCÍA, Pima County Department of Health
J. NADINE GRACIA, Office of Minority Health, U.S. Department of Health and Human Services
JEFFREY A. HENDERSON, Black Hills Center for American Indian Health
EVE J. HIGGINBOTHAM, University of Pennsylvania
CARA V. JAMES, Centers for Medicare & Medicaid Services
MELENIE MAGNOTTA, Aetna Foundation
OCTAVIO N. MARTINEZ, JR., Hogg Foundation for Mental Health, and The University of Texas at Austin
NEWELL E. McELWEE, Merck & Co., Inc.
PHYLLIS D. MEADOWS, The Kresge Foundation
CHRISTINE RAMEY, Health Resources and Services Administration
MELISSA A. SIMON, Northwestern University Feinberg School of Medicine
PATTIE TUCKER, Centers for Disease Control and Prevention
UCHENNA S. UCHENDU, Office of Health Equity, U.S. Department of Veterans Affairs
ROHIT VARMA, University of Southern California
WINSTON F. WONG, Kaiser Permanente

Roundtable on the Promotion of Health Equity and the Elimination of Health Disparities Staff

KAREN M. ANDERSON, Director, Roundtable on the Promotion of Health Equity and the Elimination of Health Disparities
ANNA MARTIN, Senior Program Assistant
ROSE MARIE MARTINEZ, Director, Board on Population Health and Public Health Practice
KENDALL CAMPBELL, 2014-2016 Puffer/ABFM Fellow at the National Academy of Medicine, and Brody School of Medicine, East Carolina University

---

[1] The National Academies of Sciences, Engineering, and Medicine's forums and roundtables do not issue, review, or approve individual documents. The responsibility for the published Proceedings of a Workshop rests with the workshop rapporteurs and the institution.

Reviewers

This Proceedings of a Workshop was reviewed in draft form by individuals chosen for their diverse perspectives and technical expertise. The purpose of this independent review is to provide candid and critical comments that will assist the National Academies of Sciences, Engineering, and Medicine in making each published proceedings as sound as possible and to ensure that it meets institutional standards for objectivity, evidence, and responsiveness to the charge. The review comments and draft manuscript remain confidential to protect the integrity of the process.

We wish to thank the following individuals for their review of this proceedings:

KENDALL M. CAMPBELL, Brody School of Medicine, East Carolina University
PAULA CARDER, Portland State University
OCTAVIO N. MARTINEZ, Hogg Foundation for Mental Health
PHYLLIS MEADOWS, The Kresge Foundation

Although the reviewers listed above provided many constructive comments and suggestions, they were not asked to endorse the content of the proceedings nor did they see the final draft before its release. The review of this proceedings was overseen by **LINDA A. McCAULEY,** Emory University. She was responsible for making certain that an independent examination of this proceedings was carried out in accordance with the standards of the National Academies and that all review comments were carefully considered. Responsibility for the final content rests entirely with the rapporteurs and the National Academies.

Contents

1 INTRODUCTION   1
   Opening Remarks, 3
     Teresa Lee
   Organization of the Proceedings, 4

2 KEYNOTE PRESENTATIONS   5
   Housing as a Linchpin of Well-Being, 5
     Lisa Marsh Ryerson
   The Five Dimensions of Housing Policy, 10
     Erika Poethig
   Discussion, 14

3 AFFORDABILITY OF HOUSING THAT SUPPORTS HEALTH
   AND INDEPENDENCE FOR VULNERABLE OLDER ADULTS
   AND INDIVIDUALS WITH DISABILITIES   17
   Financial Security and Housing for Adults with Disabilities, 17
     Purvi Sevak
   Projections for Housing a Growing Elderly Population, 21
     Jen Molinksy
   Discussion, 28

4 DESIGN FEATURES OF ACCESSIBLE HOUSING FOR OLDER
   ADULTS AND INDIVIDUALS WITH DISABILITIES 31
   Life Starts at Home: Linking Home Environment and Quality of
   Life for People with Disabilities, 32
     Bryce Ward
   Habitat for Humanity: Helping People Age in Place, 35
     Corneil Montgomery
   The Vermont Center for Independent Living:
   Improving Home Access, 38
     Patricia Tedesco
   Discussion, 40

5 MODELS CONNECTING AFFORDABLE HOUSING
   AND SERVICES AS A PLATFORM FOR HEALTH AND
   INDEPENDENCE 45
   Supportive Housing to Improve Health, 46
     Peggy Bailey
   U.S. Department of Housing and Urban Development Section 811
   Project Rental Assistance Program, 48
     Katina Washington and Lisa Sloane
   Community Aging in Place—Advancing Better Living for Elders, 52
     Sarah Szanton
   Support And Services at Home, 56
     Molly Dugan
   Promising Models in Oregon, 59
     Rose Englert
   Moderated Discussion with Panel Speakers, 61
   Discussion, 64

6 REACTORS PANEL ON POLICY IMPLICATIONS AND
   RESEARCH NEEDS 67
   Reactor Comments, 67
     Dara Baldwin
     Anand Parekh
     Robyn Stone
     Uchenna S. Uchendu
   Discussion and Closing Remarks, 74

REFERENCES 77

APPENDIXES
A Workshop Agenda 81
B Biographical Sketches of Workshop Speakers and Reactors 87

1

Introduction[1]

Accessible and affordable housing can enable community living,[2] maximize independence, and promote health for vulnerable populations. However, the United States faces a shortage of affordable and accessible housing for vulnerable low-income older adults and individuals living with disabilities. This shortage is expected to grow over the coming years given the population shifts leading to greater numbers of older adults and of individuals living with disabilities.

Housing is a social determinant of health and has direct effects on health outcomes, but this relationship has not been thoroughly investigated. To better understand the importance of affordable and accessible housing for older adults and people with disabilities, the barriers to providing this housing, the design principles for making housing accessible for these individuals, and the features of programs and policies that successfully provide affordable and accessible housing that supports community living for older adults and people with disabilities, the Health and

---

[1] The planning committee's role was limited to planning the workshop, and this Proceedings of a Workshop has been prepared by the workshop rapporteurs as a factual summary of what occurred at the workshop. Statements, recommendations, and opinions expressed are those of individual presenters and participants, and are not necessarily endorsed or verified by the National Academies of Sciences, Engineering, and Medicine, and they should not be construed as reflecting any group consensus.

[2] For the purposes of this Proceedings of a Workshop, "community living" refers to individuals being able to live at home in their communities as opposed to living in a residential facility, unless stated otherwise in the text.

> **BOX 1-1**
> **Statement of Task**
>
> An ad hoc committee will plan, organize, and implement a 1-day public workshop to explore the role of housing as a social determinant of health for older adults and people with disabilities, particularly among people of color and low-income groups. Accessible, affordable housing can be a platform for enabling community living, maximizing independence, and promoting positive health outcomes for vulnerable populations. The affordability and availability of this housing, however, is often a major barrier.
>
> The role of housing as a social determinant of health is under-investigated. The workshop will feature invited presentations and discussions focused on topics surrounding the affordability, availability, and importance of supportive housing for these populations, including housing supply, supportive housing needs, differences between rural and urban settings, social isolation, subsidies, and opportunities for policy changes.
>
> The committee will plan and organize the workshop, develop the agenda for the workshop, select and invite speakers and discussants, and moderate or identify moderators for the discussions. A single summary of the presentations and discussions at the workshop will be prepared by a designated rapporteur in accordance with institutional guidelines.

Medicine Division and the Division of Behavioral and Social Sciences and Education, both of the National Academies of Sciences, Engineering, and Medicine (the National Academies), with support from a group of sponsors (see page ii for a list), jointly convened a public workshop on December 12, 2016, in Washington, DC. The Forum on Aging, Disability, and Independence[3] (the forum) and the Roundtable on the Promotion of Health Equity and the Elimination of Health Disparities[4] (the roundtable) hosted the workshop. The forum meets to discuss how to support independence and community living for people with disabilities and older adults. The roundtable promotes health equity and the elimination of health disparities by advancing the visibility and understanding of the inequities in health and health care among racial and ethnic populations; by amplifying research, policy, and community-centered programs; and by catalyzing the emergence of new leaders, partners, and stakeholders.

An ad hoc committee (see Box 1-1 for the planning committee's state-

---

[3] For more information, see http://www.nationalacademies.org/ADIForum (accessed May 11, 2017).

[4] For more information, see http://www.nationalacademies.org/hmd/activities/selectpops/healthdisparities.aspx (accessed May 11, 2017).

ment of task) planned and designed the workshop to meet the following objectives:

- Summarize the knowledge on housing as a social determinant of health and as a platform for health and independence for vulnerable older adults and individuals with disabilities.
- Highlight successful and promising collaborations for providing affordable and accessible housing, with consideration for differences between rural, suburban, and urban settings.
- Explore sustainable and scalable strategies, policies, and practices to support linking affordable housing with services to benefit health and optimize independence.
- Discuss data needs and research gaps to measure the effectiveness of models of housing with supportive services for vulnerable older adults and individuals with disabilities.

Under the National Academies guidelines, workshops are designed as convening activities and do not result in any formal findings, conclusions, or recommendations. Furthermore, the Proceedings of a Workshop reflects what transpired at the workshop and does not present any consensus views of the planning committee or workshop participants. The purpose of this proceedings is to capture important points raised by the individual speakers and workshop participants. Speaker presentations slides are also available on the workshop website.[5]

OPENING REMARKS

In her introductory comments at the workshop, Teresa Lee, the executive director of the Alliance for Home Health Quality and Innovation, noted that housing is clearly a significant social determinant of health, particularly in the Medicare home health context. "It is one of those issues that is often too big for us to even think straight about," she said, "so I am excited that today we are going to have a chance to delve in and think hard and long about how housing has an effect on health care and what we can do in concrete terms."

Lee also pointed out that there is no one-size-fits-all solution to making housing more affordable and accessible, and she predicted that the day's presentations would reflect that by describing what strategies are working well in different parts of the United States. Lee said that her hope was that the examples provided at the workshop would trigger discus-

---

[5] See http://www.nationalacademies.org/hmd/Activities/Aging/AgingDisability Forum/2016-DEC-12.aspx (accessed January 18, 2017).

sions about how to apply and adapt promising programs to other regions of the country and about how to scale these programs to meet the needs of more people, particularly in underserved communities. She also predicted that a common theme across the presentations would be the crucial role that partnerships at the federal, state, and local levels can play in creating successful programs.

ORGANIZATION OF THE PROCEEDINGS

An independent planning committee (see page v for the list of committee members) organized the workshop (see Appendix A for the agenda) in accordance with the procedures of the National Academies. This publication describes the presentations and discussions that occurred during the workshop. Generally, each speaker's presentation is reported in a section attributed to that individual. Chapter 2 recaps the two keynote presentations which provided a foundation for the remainder of the workshop's discussions. Chapter 3 examines issues regarding the affordability of and financing for housing that promotes health and independence among vulnerable older adults and people with disabilities; Chapter 4 considers some of the design features that make housing accessible for these populations; and Chapter 5 describes six models that use affordable and accessible housing as platforms for health and independence. Chapter 6 discusses the potential policies and research needed to support efforts to increase the supply of affordable and accessible housing for vulnerable older adults and individuals with disabilities.

In accordance with the policies of the National Academies, the workshop attendees did not attempt to establish any conclusions or recommendations about needs and future directions, focusing instead on issues identified by the speakers and workshop participants. In addition, the planning committee's role was limited to planning the workshop. Workshop rapporteurs Joe Alper, Karen Anderson, and Sarah Domnitz prepared this Proceedings of a Workshop as a factual summary of what occurred at the workshop.

2

Keynote Presentations

To provide a foundation for the workshop's discussions, two keynote speakers discussed the health consequences that inadequate housing can have for vulnerable older adults and individuals with disabilities and addressed some of the barriers that constrain the supply of such housing in the face of rising demand. Lisa Marsh Ryerson, president of AARP Foundation, described housing as the linchpin of well-being and discussed some of the programs her organization has supported in several communities. Erika Poethig, a fellow and the director of urban policy initiatives at the Urban Institute, then described the five dimensions of housing policy along with potential areas for changing housing policy to better support the needs of older adults and individuals with disabilities. A brief discussion moderated by Teresa Lee of the Alliance for Home Health Quality and Innovation followed the two keynote presentations.

HOUSING AS A LINCHPIN OF WELL-BEING

Lisa Marsh Ryerson
President
AARP Foundation

The definition of a linchpin is a person or thing that holds something together, said Lisa Ryerson of AARP Foundation. At AARP Foundation, Ryerson said, she and her colleagues view housing as the linchpin of well-being both for individuals in a community and for the community itself. With affordable, livable, and healthy housing, people of all ages and abilities have the opportunity to thrive, she said. Housing and the location of housing are social determinants of health that affect whether an individual has access to sustaining and critical services, to transportation to those services, and to important social connections. Those who struggle to pay their rent or mortgage face critical decisions on a daily basis about what other essentials they will not be able to pay for, such as food, medicine, health care, and transportation. In addition, Ryerson said, people whose housing circumstances contribute to falls, who live in unsafe conditions, or who experience social isolation often find that their health is impaired in the short and long term.

Currently, Ryerson said, approximately 1 percent of the nation's housing stock is adequately equipped to meet the needs of older adults, having such essential elements as no-step entryways or grab bars (Baker et al., 2014). "That calls us to find long-term solutions to improve homes and communities and to provide health services so that older adults do not have to leave their homes or leave their communities when their health status or health needs begin to change," Ryerson said. "When housing is at risk, overall well-being is then threatened." Seeking to play a role in developing and implementing those solutions, AARP Foundation works to help ensure that vulnerable older adults with low incomes have access to healthy and nutritious food and to safe and affordable housing, that they are able to maintain critical social connections, and that they have opportunities to generate income.

Collaboration is essential to AARP Foundation's work, Ryerson said. "It fuels the collective impact that makes a real difference in people's lives." In order to have a positive collective impact, institutional egos must be put aside, with all partners understanding that no single policy, program, or group can solve a complex social problem by itself. "It gets us out of our silos—the comfortable compartments that can restrict our thinking and limit our capacity—to join with others to solve these problems," she said. Collaboration is essential to tackling persistent, stubborn, and complex social problems such as the availability of affordable and accessible housing that can enable older adults and those with disabilities to continue living in their communities. "Collective impact and a commitment to it allow us to bring a wide range of actors, each with a different lens, background, and expertise, but each sharing a commitment to a common goal," Ryerson said. "This is how we can meet the needs of all people, including communities of color, the LGBTQ[1] community, and people with disabilities."

To put the issue concerning older adults in context, Ryerson explained

---

[1] Lesbian, gay, bisexual, transgender, and queer or questioning.

that some 10,000 people in the United States reach age 65 every day. The question for many of these Americans is whether their expected quality of life will match their hopes for their future health and well-being. With more than 19 million older adults living in unaffordable or inadequate housing, and given the projections of a dramatic increase in the older adult population in the coming decades, solving this problem becomes even more urgent. For example, falls are the leading cause of injury and injury-related death among adults 65 and older, and according to the National Institute on Aging, 6 of every 10 falls happen at home (NIA, 2013).

The question, Ryerson said, is whether the nation is meeting the housing needs for older adults, and according to a study on older adults and housing conducted by Harvard University, the answer is clearly "No" (Baker et al., 2014). "This disturbing reality drives us to invest in innovative solutions to help people age in place, as almost 90 percent of older adults want to do," she said. "For older adults to thrive and to live with independence and dignity in the least restrictive setting, we need new ways to provide health services in the most conducive settings, and this means looking beyond the doctor's office." It is important, she added, to embrace innovation through a more integrated, person-centered model of health care delivery that broadens the concept of health care so that it is not restricted to a hierarchy headed solely by physicians and so that care can be delivered in more places.

As an example of innovative thinking, Ryerson described a program called Care Angel—winner of AARP Foundation's Aging in Place challenge—which is using an artificial intelligence-driven assistant that calls older individuals daily to check on their well-being. This assistant records how the older adults are doing and whether or not they have taken their medication. It also asks about their appetite, sleep quality, blood pressure, glucose levels, and other health-related questions. "Care Angel is a compelling example of how we can bring seemingly incongruent sectors together to achieve a common goal, enabling an older person to stay at home, and connecting them potentially to vital health services," Ryerson said.

Although technology can provide new solutions with great potential to link housing and health, technology cannot be the only answer because human connection is also vital to well-being and quality of life, Ryerson said. The Community Aging in Place—Advancing Better Living for Elders (CAPABLE) program[2] developed at the Johns Hopkins School of Nursing, for example, is using teams of occupational therapists, registered nurses,

---

[2] See http://nursing.jhu.edu/faculty_research/research/projects/capable (accessed February 8, 2017).

and licensed contractors to go into homes and work with older adults to make crucial adjustments in their homes and in their daily routines to reduce the likelihood of hospitalization and the need for nursing home care (see Chapter 5). The Green House Project[3] (Miller et al., 2016; Sharkey et al., 2011) is another program that links housing and health by providing skilled, individualized nursing care in an environment that looks and feels like home. This latter project is one of several that receives funding from Age Strong, an impact investment initiative of AARP, AARP Foundation, the Calvert Foundation, and Capital Impact Partners that leverages private capital investment for social good.

Simply knowing that affordable and accessible housing is central to both health and well-being—and to achieving the Triple Aim of better quality of care, better outcomes, and lower costs[4]—is not the same as doing something about affordable and accessible housing, Ryerson said. Despite growing interest in collaborations to bring housing and health together, she said, these types of partnerships have been slow to take off, slow to be replicated, and slow to be taken to a larger scale. During a meeting held by AARP Foundation and LeadingAge, four key elements were identified as critical for collaborations between affordable housing and health care to take hold:

1. Identify the specific population to be served.
2. Develop a partnership model for coordinating and integrating services.
3. Share information between housing and health care partners.
4. Create a sustainable business model for housing and health programs.

The focus of AARP Foundation's work, Ryerson said, is human-centered design. "We look at both traditional and nontraditional models to lift up efforts that address social determinants of health, and regardless of the model, the common goal is to help strengthen the community and to help older adults thrive." To carry out this effort, AARP Foundation looks for individuals and community-based organizations and professions across sectors that are primed to help and ready to channel their energy for good in new and perhaps unexpected directions. These partners could be organizations that are nonprofit, for-profit, public, private, or academic. For example, Ryerson said, the ShopRite supermarket chain

---

[3] For more information, see http://www.thegreenhouseproject.org (accessed February 8, 2017).

[4] For more information about the Triple Aim, see http://www.ihi.org/Engage/Initiatives/TripleAim/Pages/default.aspx (accessed February 27, 2017).

found an opportunity to help low-income communities in Philadelphia that went beyond providing healthy food at affordable prices. Ryerson explained that the company provides access to health services, benefits programs, and nutrition services, including dietitians and nutritionists, within its stores. The nutritionists and dietitians conduct cooking demonstrations and workshops and provide one-on-one advice. They also take biometric readings for care-managed patients and track prescribed diet adherence and health changes. Furthermore, the stores provide transportation to and from the home and free food delivery for homebound individuals because the company recognized that these additional services would only be valuable if people could access them. "The need to secure, maintain, and sustain social bonds in order to generate good health over time is equally important," Ryerson said. Perhaps not surprisingly, these stores have become community centers. These stores, she added, are not only transforming "food deserts into wellsprings of good nutrition, but they're also building stronger communities by acting as a critical connector to services and to benefits."

Going forward, it will be important that the health care policy debate in Washington, DC, and in communities around the nation is informed by the lessons learned from promising models for creating partnerships between affordable housing and health care to serve older adults, Ryerson said. "Why not seize this moment in time to find the nexus between seemingly unrelated issues, and for us, as a learning community, to bring those to the fore?" she asked. "It is time for groups that have had little experience working with each other to move from merely sharing an interest to coming forward in true collaboration." She called on the philanthropic community, the research community, and the government to hasten this movement by supporting and analyzing promising partnerships in the field. "Let us use the skills, experience, empathy, and proximity of people who are well positioned to make a positive impact in a . . . place that falls outside the norm, and let us identify and deploy what I call agents of opportunity, such as Care Angel and other innovators, to help seniors age independently at home and in community."

"Above all," said Ryerson in closing, "let us stop tackling the issues of housing and health on parallel tracks that do not intersect. Instead of talking about health and housing separately, let us do what we are doing today and advance the conversation about housing and health together, a conversation that includes practical ideas for addressing inadequate housing stock and quality as well as disparities based on age, race, ethnicity, gender identity, sexual orientation, and disability."

THE FIVE DIMENSIONS OF HOUSING POLICY

Erika Poethig
Fellow and Director of Urban Policy Initiatives
Urban Institute

Erika Poethig of the Urban Institute began her remarks by agreeing with Ryerson that breaking down silos is essential if the nation is going to provide affordable and accessible housing for those who need it. Poethig said that during her career focused on busting silos she has learned to "never assume that your framework is the framework that the other person across the table shares."

Poethig said that she and her colleagues generally frame housing policy using five dimensions: quality, affordability, tenure (i.e., the difference between renting and owning a home), stability, and location. "Each of these, I would argue, has a differential impact on health outcomes, especially for older adults and people with disabilities," she said.

Housing *quality*, Poethig said, first became an impetus for policy change early in the 20th century. Although quality drew less attention later in the 20th century, more recently it has re-emerged as an area of policy focus because of elements that can especially affect older adults and people with disabilities. For example, homes that have not been retrofitted, that have poor lighting, or that do not have a bathroom on the first floor are the types of quality conditions that can lead to falls. Not only are falls in the home a major cause of morbidity and mortality in older adults, but in 2015 falls were estimated to account for $31 billion in direct medical costs to Medicare.

The *affordability* of housing can have implications for an individual's health. If housing is unaffordable, it compromises an individual's ability to afford food, medication, and services that are important for maintaining health. This is especially true for many older adults and people with disabilities who support themselves on fixed incomes.

Housing *tenure* refers to renting versus owning a home. Not having the financial freedom to choose between renting or owning, or not having sufficient rental or purchasing options that are suitable for older adults or people with disabilities, can be another constraint on the ability to access needed services in some communities, Poethig said.

Housing *stability* refers to how often an individual has to move between homes. Moving can be stressful for anyone, Poethig said, even when an individual is moving to a more positive or better setting. Forced and frequent moves that disrupt care access can be particularly damaging to health. The recent foreclosure crisis, she said, had a negative effect on housing stability, affecting not only those who owned their homes but also those who were renting a home that was foreclosed on.

Housing *location*, Poethig continued, is often characterized as being a construct designed to support opportunities for families and young children. Poethig said that supporting opportunity in terms of location for older adults and individuals with disabilities is also important and often not discussed in conversations about housing location.

Interplay Among the Dimensions of Housing Policy

The five dimensions of housing are interrelated, Poethig said, and they can sometimes be in tension with one another. Housing affordability, an area in which the federal government plays a central role, became a major emphasis of federal housing policy in the late 1960s, she said. Today, 83 percent of the U.S. Department of Housing and Urban Development's (HUD's) budget is dedicated to programs that provide federal rental assistance. Nonetheless, Poethig said, some 11.2 million households face severe rent burdens, as defined by paying more than half of household income on rent, and 91 percent of these severe-rent-burden households are of extremely low income, meaning they earn less than 30 percent of the area median income. Some 16 percent of the households facing both severe rent burdens and extremely low income are older adults, and 26 percent include at least one adult with a disability. Contrary to popular belief, the problem of severe cost burden is not confined to big cities, Poethig said—housing affordability is an issue that touches suburban, urban, and rural communities.

Federal housing assistance is distributed by a lottery system, Poethig explained. This means that only approximately one in four households that is eligible for assistance receives it. "That is a major challenge in our policy framework," she said, adding that the aging baby boomer population is going to put even more pressure on rents over time. The Urban Institute has projected that over the next 15 years there will be 13 million new renter households and the number of older adult renter households will more than double, from 5.8 million to 12.2 million. A study conducted by HUD and published in 2015 estimated that only approximately 3–4 percent of households with at least one person with a disability received a housing unit with features designed for individuals with mobility disabilities, even though some 20 percent of the U.S. adult population has at least one disability. Thus, not only is there a constraint in terms of affordability, Poethig said, but there is also a constraint in terms of the supply of housing that is of a quality that is suitable for people with mobility disabilities.

One might ask, Poethig said, why the market is not responding to this demand by building more homes. The answer, she said, lies partly in the fact that the nation's housing supply overall is already falling short of the demand. However, she added, the construction of affordable housing will continue to lag because development costs—e.g., construction, labor, and land—are too expensive to produce affordable housing for those living on a fixed income. Therefore, subsidies will always be needed to fill the gap between what it costs to build housing and what many individuals can afford to pay.

A study conducted by Urban Institute found that older adults in the United States spend much more of their income on housing than on health, with housing accounting for approximately 25 percent of their income, Poethig said (Johnson, 2015). Due to property taxes and the costs associated with maintaining a home, this holds true even among those who own their home free and clear, she said. For the 7 million older adults whose incomes are 125 percent of the poverty level, housing costs account for 74 percent of their income. "So absent increased resources for federal rental assistance, America's older adult population and people with disabilities will continue to face these particular housing instability challenges, and I would argue, poor health outcomes," Poethig said. One policy barrier to addressing this problem is what Poethig called the "wrong pockets problem," which refers to the inability of the housing sector to capture the savings to health care expenditures that could result from an investment in housing. "We are just going to see these costs land in another part of the ledger," she said.

One concern regarding stability is the effect that foreclosures have had on people with disabilities living in group homes established after the 1999 Olmstead Decision ruling by the U.S. Supreme Court.[5] Poethig said that HUD does not have any mechanism to address foreclosures or the displacement and housing instability that can result from them. Poethig suggested that the U.S. Department of Health and Human Services provide more clarity on what the technical details of the ruling require in order to identify other possible options for the delivery of services and for arranging housing in a community. This clarity is needed both for the people who work at HUD to develop programs to support the law and for those who work at the U.S. Department of Justice to enforce the law.

HUD does have resources through the Community Development Block Grant Entitlement Program[6] to fund repairs, improvements, and retrofits that can benefit homeowners who are older or who have dis-

---

[5] *Olmstead v. L.C.*, 527 U.S. 581 (1999). This ruling requires that community-based services be provided to individuals with disabilities when appropriate.
[6] For more information, see https://www.hudexchange.info/programs/cdbg-entitlement (accessed February 14, 2017).

abilities, which can help them to remain in their homes. However, Poethig said, these resources are not sufficient to meet demand and are not dedicated solely for retrofitting homes. She noted that the Bipartisan Policy Center[7] has been developing ideas for securing new sources of funding for retrofitting homes.

Identifying Policy Levers

Approximately half of the people receiving federal rental assistance are older adults and people with disabilities, but assistance for families with children is equally important, Poethig said. She said that she worries that advocates for these groups will be in tension with one another about the positive role that federal rental assistance can play in improving life outcomes. To avoid this, Poethig suggested that federal rental assistance be expanded to all who are eligible, although she admitted that this is unlikely to happen in the current political climate. Equally important, she added, is preserving the assets that are in place today. Private owners have opted out of their federal subsidy contracts in many communities across the United States. It is destabilizing to older adults and can force them into having to move into another community, she said. In some communities, she added, regulatory barriers can also reduce the supply of affordable and accessible housing.

A new model called "pay for success" is an evidence-based policy tool that could be useful for financing affordable and accessible housing and also for increasing the understanding of the connection between housing and health, Poethig said. This financing model is designed to relieve the government of the risk of investing in a new program. It combines private capital and private investors to provide a source of funds to support the scaling up of evidence-based social programs, and the government repays the investors if the program is successful. The concept of partnerships between different sectors is at the very heart of this program. Poethig said that evidence-based supportive housing[8] models are getting attention as potential candidates for pay-for-success programs. This model is still new, with only 11 of the financing programs having begun at this point. One example is found in Denver, Colorado, which is using pay-for-success to pay for supportive housing services, with other programs subsidizing the housing itself. Poethig noted that the recently enacted Comprehensive

---

[7] For more information, see http://bipartisanpolicy.org (accessed February 14, 2017).
[8] Supportive housing is non-time-limited affordable housing matched with voluntary ongoing supportive services appropriate to the tenant.

Addiction and Recovery Act[9] had originally included $100 million to support the guarantees of pay-for-success programs, but this provision was pulled from the bill before it became law. Urban Institute, she added, has a pay-for-success initiative.[10]

Expanding the Evidence Base

Poethig concluded her presentation with a call to increase efforts to conduct research and collect data to better understand successful partnerships between housing and health. She cited CareOregon[11] and the Support And Services at Home (SASH) program in Vermont[12] as successful programs from which more can be learned through research and data collection in order to identify the benefits of connecting housing and health (see Chapter 5). "We need to continue to invest in the evidence to build our understanding of how these kinds of partnerships work, what costs they may be saving, and how we can actually cover those costs," she said. "That will help us in communication with our partners in Congress." Poethig also called for more work to disaggregate data or, in some cases, to match data from different agencies in order to increase understanding and further fine-tune available programs.

DISCUSSION

Dara Baldwin of the National Disability Rights Network started the discussion by commenting that even though rental assistance programs are earmarked for various groups, older adults, people with disabilities, and families are all interconnected: families may have parents or children with disabilities, for example. Baldwin noted the need for data on whom the rental assistance is helping. Poethig agreed that this is an important nuance and an intersection where different communities can come together to advocate for more rental assistance.

Margaret Campbell of Campbell & Associates commented that in addition to quality, accessibility should also be considered a critical dimension of housing policy. She said that HUD does not have standards for accessibility and suggested that the agency develop standards for accessibility similar to those that it has for affordability. Poethig agreed that HUD should create standards for accessibility, adding that she would

---

[9] Comprehensive Addiction and Recovery Act of 2016, Public Law 114-198, 114th Cong. (July 22, 2016).
[10] For more information, see http://pfs.urban.org (accessed February 9, 2017).
[11] For more information, see http://www.careoregon.org (accessed February 14, 2017).
[12] For more information, see http://sashvt.org (accessed February 14, 2017).

also like to see the agency update its quality standards to include more considerations for accessibility because those are the standards by which it assesses housing units.

Daniel Davis of the Administration for Community Living asked the panelists if they have insights into why builders are not building housing to the accessibility standards that people need. Ryerson said that this problem confounds her, too, especially considering that accessibility increases livability for everyone across the age span living in an accessible home. Perceived cost is a factor, she acknowledged, which makes it important to look at the evidence that shows that meeting accessibility standards comes with overall cost savings. One aspect of the problem, Poethig said, is that the housing in many parts of the country was built before creation of the standards set by the Americans with Disabilities Act.[13] Another issue, she added, is that 50 percent of the nation's rental housing is in small buildings such as single-family detached units whose owners may not be aware of the rules and standards as they apply to the single-family housing market. Daniela Koci of Loveland Village said that her organization has developed an affordable housing apartment complex for individuals with developmental disabilities and their caregivers. A big hurdle the organization faced was that the standards for cost per unit set by state and federal funders such as HUD were lower than the thresholds for many universal design principles, which created challenges for procuring funding. "So if you are a private developer, it is not worth it on the profit side, and if you are a nonprofit it is hard to secure funding because the cost per unit is so high up front," she said.

Caroline Blakely of Rebuilding Together asked the keynote speakers for suggestions on how to effectively communicate the importance of the connection between housing and health to leaders in the private capital and corporate sectors. Ryerson said that it is important to generate and disseminate data to show how safe, adequate housing benefits both individual health and community health. Lee agreed that building an evidence base is crucial for bringing new partners to the table, including those in the for-profit world. She added that some of the new health care delivery and payment models might provide an incentive for more activity in the housing realm. She pointed to recent work by Eric Fisher and colleagues at the Dartmouth Institute for Health Policy and Clinical Practice (Fraze et al., 2016), who found that leaders of accountable care organizations are starting to think about the connections between health and social determinants of health, such as housing, transportation, and food. "I think we might be getting close to a tipping point of connectedness, but we are not quite there yet," Lee said. Poethig added that the

---

[13] Americans with Disabilities Act of 1990, Public Law 101-336, 101st Cong. (July 26, 1990).

30-day hospital readmission metric for Medicare reimbursement could serve as a lever to convince health care systems to pay more attention to housing issues.

Sarah Triano of Centene Corp., a Medicaid managed care organization, asked the panelists if they knew of any innovative projects using telehealth. Lee replied that telehealth is used often in the home health context, in part because of the payment structure—i.e., providers are paid per episode, not per visit. Her organization's website[14] includes innovation profiles that it has developed involving telehealth. This technology, Lee said, may help a strained workforce meet the growing needs of older adults and individuals with disabilities who want to remain in their homes.

Karen Anderson of the National Academies of Sciences, Engineering, and Medicine asked the panelists to comment on how the issues and ideas they raised specifically affect minorities and low-income individuals. Poethig replied that HUD's federal rental assistance program primarily serves people who are very poor and vulnerable, and minorities represent a disproportionate percentage of those receiving rental assistance. Therefore, she said, federal rental assistance presents an opportunity and a platform upon which to connect and address broader issues surrounding health disparities. Ryerson agreed and noted that it is important when forming solutions to be thoughtful about how different lenses of identity might intersect. "Communities cannot and will not be healthy unless we make that consideration at the forefront of all of our design thinking," she said. "We have to stop this collective thinking that it is okay to subtract the talent and contributions of some members in communities."

---

[14] For more information, see http://www.ahhqi.org (accessed February 17, 2017).

3

Affordability of Housing That Supports Health and Independence for Vulnerable Older Adults and Individuals with Disabilities

The workshop's first panel featured two presentations addressing the affordability of housing. Purvi Sevak, a senior researcher at Mathematica Policy Research and a professor of economics at Hunter College of the City University of New York, spoke about the ties between financial security and housing and the importance of infusing disability awareness into all housing policy discussions. Jen Molinsky, a senior research associate at the Joint Center for Housing Studies of Harvard University, then discussed the results of a new report from the Joint Center on the implications for housing based on the growing population of older adults. A discussion with the workshop audience followed the two presentations.

FINANCIAL SECURITY AND HOUSING FOR ADULTS WITH DISABILITIES

Purvi Sevak
Senior Researcher, Mathematica Policy Research
Professor of Economics,
Hunter College of the City University of New York

Disability, said Purvi Sevak of Mathematica Policy Research and Hunter College of the City University of New York, is a function of both health and the environment. Disability is not a yes-or-no concept in that environmental factors, including housing, can affect functioning and thus turn a health problem into a disability. Therefore, she said, it is important to keep disability in mind when making housing policy.

Some 10.5 percent of all adults ages 18 to 64—20 million nonelderly individuals in the United States—have disabilities, with the prevalence of disability higher among vulnerable populations. For example, Sevak said, 20 percent of individuals living in poverty have a disability, as do 27 percent of individuals living in public housing, 34 percent of individuals with subsidized rent, and 40 percent of homeless individuals (Hoffman and Livermore, 2012; Houtenville et al., 2015). Because disability is so prevalent in vulnerable populations, she said, any discussion about housing policies for these populations is not complete without a discussion of disability. Sevak also noted that in the context of the need for accessible housing, it is irrelevant whether disability leads to financial insecurity or whether financial insecurity increases the likelihood of developing health impairments that lead to disability. In either case, accessible housing is the focus, although the direction of the cause-and-effect relationship may be important to discussions about reducing the rates of disability and improving economic well-being.

There are many different types of disability, Sevak noted, so individuals with disabilities may have very different housing needs. Federal surveys, such as the American Community Survey,[1] classify disability into six categories: hearing, vision, cognition, ambulation, self-care, and independent living (see Figure 3-1). Sevak said that in discussions about housing, the emphasis is usually on individuals who have an ambulatory disability, which is present in slightly more than 5 percent of the adult population ages 18 to 64 years. Concern is likely to focus on the ramps and bathrooms fixtures needed to make federal housing accessible for an individual with an ambulatory disability, but other types of disabilities require other types of accommodations. Making housing accessible to an individual with a cognitive disability, for instance, should begin with concern about the complexity of the application forms, Sevak said

Having a disability is associated with greater financial vulnerability, Sevak said. In 2015 the employment rate among individuals without disabilities was 75.4 percent versus 34.4 percent for those with a disability, even though the majority of individuals with disabilities report in surveys that they would like to work. Median earnings among individuals with disabilities who were working was $21,232 versus $31,324 for individuals who were working but did not have a disability. Median family income was $36,800 for the household of an adult with a disability versus $66,000 for households where the adults did not have a disability. The poverty

---

[1] For more information, see https://www.census.gov/programs-surveys/acs (accessed February 22, 2017).

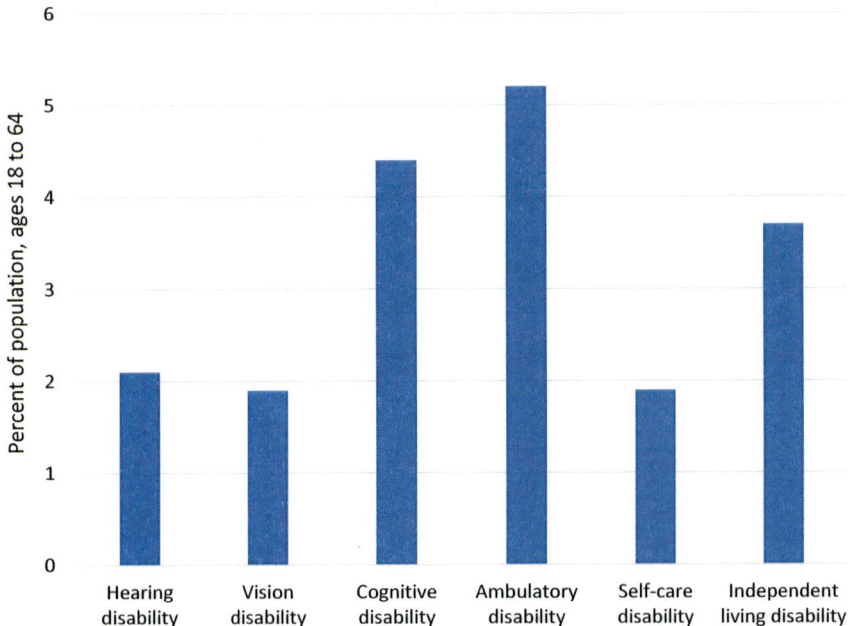

FIGURE 3-1 Percentage of adults ages 18 to 64 years who have different types of disabilities.
SOURCES: Sevak presentation, December 12, 2016. Data from Houtenville et al., 2015.

rate among those with disabilities was nearly double that of those without disabilities in 2015, and only 54.5 percent of those with disabilities lived in an owned home, compared with 61.7 percent of individuals without disabilities. Sevak noted that many individuals with disabilities collect benefits through either the Social Security Disability Insurance program or the Supplemental Security Income program. Once individuals with disabilities reach age 65, their income is less than one-third of the income of their peers who do not have a disability ($24,900 versus $75,900), which puts the financial security of those with disabilities at greater risk than that of their peers as they enter retirement. Furthermore, among adults ages 18 to 64 years who have incomes below the poverty level, those who have a disability are more likely to have material hardship (She and Livermore, 2007) (see Figure 3-2). These disparities persist in households with income above the poverty level as well, Sevak said.

Given these data, Sevak said, it should not be a surprise that disability

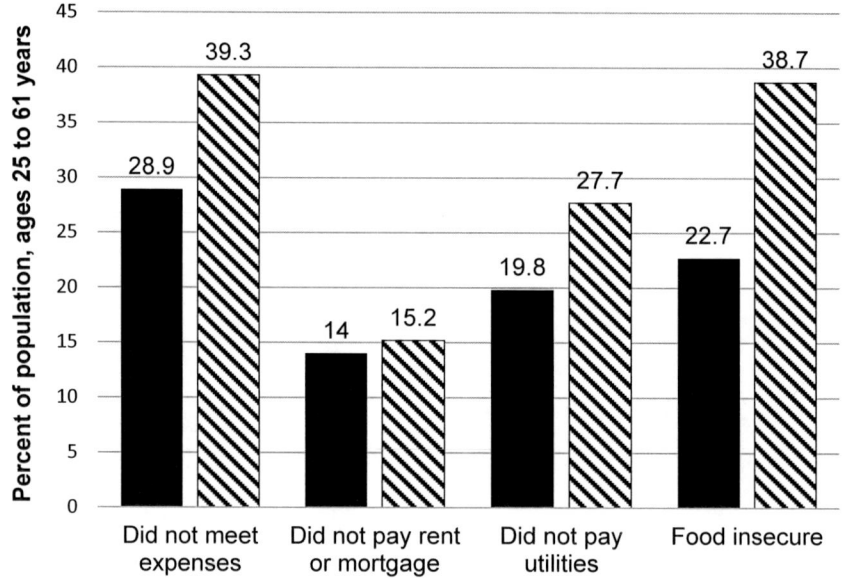

FIGURE 3-2 Disability is associated with greater rates of material hardship.
NOTE: FPL = federal poverty level.
SOURCES: Sevak presentation, December 12, 2016. Data from She and Livermore, 2007.

is also associated with disparities in housing. Data from the American Housing Survey have shown that 48 percent of people with disabilities, compared with 38 percent of those without disabilities, report housing deficiencies which include holes in the floor, rodents, leaks, toilet breakdowns, unsafe drinking water, cracks in the home's foundation, and missing electrical outlets (Hoffman and Livermore, 2012). Similarly, she said, 76 percent of individuals with disabilities, compared with 66 percent of those without disabilities, report neighborhood problems such as crime, odors, noise, vandalism, and poor condition of neighborhood streets. These disparities persist, she added, even after controlling for demographics, income, region, and household size.

In concluding her presentation, Sevak offered two takeaway messages. The first was the importance of disability awareness in policy and practice, especially when one is focusing on vulnerable populations. Considering the needs of individuals with disabilities is paramount when making housing-related policies, she said. The second takeaway was that individuals with disabilities are a heterogeneous group and thus have varied needs with respect to housing.

PROJECTIONS FOR HOUSING A GROWING ELDERLY POPULATION

Jen Molinsky
Senior Research Associate
Joint Center for Housing Studies of Harvard University

The U.S. population is aging rapidly, said Jen Molinsky of the Joint Center for Housing Studies of Harvard University, and by 2035 there will be 79 million people ages 65 and older, compared with 48 million in 2016. This demographic shift led her and her colleagues to conduct a study to understand how the soaring population of older adults will affect the nation's housing needs over the next 20 years (Joint Center for Housing Studies, 2016). This study focused on three dimensions of housing that Molinsky said are critical for helping people remain independent and secure in their homes: accessibility, the home as a locus for long-term care, and affordability.

Projections of the Older Population and Households

Households headed by individuals age 65 and older are projected to grow from 30 million in 2016 to 41 million in 2025 and to nearly 50 million by 2035, at which point they will account for one-third of households in the United States. Households headed by someone of age 80 and older will double from 8 million in 2016 to 16 million in 2035, with much of the growth occurring between 2025 and 2035 when the leading edge of the baby boomers is in their 80s. The oldest households, Molinsky said, will drive growth in the number of single-person households over the next two decades.

Most households headed by an older adult own their homes, with more than 80 percent of Americans age 65 and older owning their home, compared with 64 percent for all U.S. households. The level of home ownership declines slightly as people enter their late 70s and early 80s and become renters, often in order to live in a home with lower maintenance demands and greater accessibility. "But looking ahead," Molinsky said, "we generally expect a high continued rate of ownership for this group." However, these projections are clouded somewhat by the uncertain long-term effects of the Great Recession, which saw individuals in their pre-retirement years experiencing a 5 percent decline in homeownership and a greater loss of wealth than adults who were age 65 and older. Among the latter, home ownership fell by only 1 percent during the Great Recession. As a result, Molinsky said, in 20 years ownership could be lower than she and her colleagues have projected.

Some older adults neither own their home nor rent, but rather live with relatives or in group quarters such as a nursing home, Molinsky said. Approximately 11 percent of adults ages 65 to 79 live in these types of arrangements, with more than half living with family, typically their children. For those 80 and older, 23 percent live in these arrangements, with nearly 12 percent living with family. In the future, the number of older adults living with family members could rise, given that living with family is more common among Hispanic and Asian households, which are demographic groups that are increasing as a percentage of the U.S. population. Currently, some 8 percent of the U.S. population ages 65 and older is Hispanic, and that is projected to rise to 13 percent by 2035, Molinsky said, while the population of older Asian adults is estimated to grow from 5 percent to 7 percent. The possibility of there being more multigenerational families in the future has some implications for such housing features as size, flexibility, and layout.

The demand for group quarters is hard to predict given that the trend over the past 20 years has shown a distinct move away from living in nursing homes even as the population over age 50 has increased, Molinsky said. The increasing availability of home-based, long-term care might continue to drive down nursing home use. However, because the population of older Americans continues to grow, the number of people living in nursing homes may still grow even if that number decreases as a percentage of the older adult population.

Estimations of Housing Accessibility

To understand how the predicted increasing prevalence of disability in the population will affect housing specifically, Molinsky and her colleagues divided disability into three categories: mobility disability; disability that affects self-care, which includes activities of daily living such as eating, bathing, dressing, and toileting; and disability that affects household activities, which includes instrumental activities of daily living such as meal preparation, food shopping, using the telephone, taking medication, driving, managing money, and housework. Their analysis found that the number of people in these three categories rises distinctly when people reach their late seventies, and that this rise occurs regardless of race, ethnicity, housing tenure, or income. "No one is immune," Molinsky said. One finding of note, she added, was that the difference in disability rates by tenure and income converge with age, while disparities by race persist (see Figure 3-3).

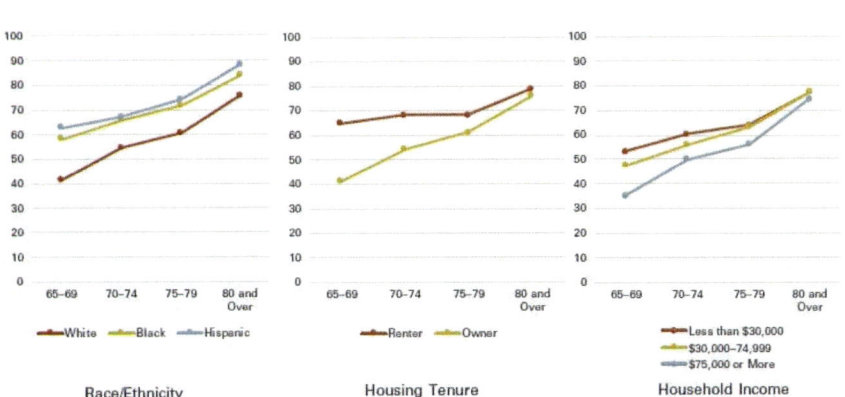

FIGURE 3-3 Differences in disability rates by race/ethnicity, tenure, and income.
SOURCES: Molinsky presentation, December 12, 2016. Joint Center for Housing Studies, *Projections & Implications for Housing a Growing Population: Older Households 2015-2035*, 2016, www.jchs.hardvard.edu. All rights reserved.

Projecting disabilities was challenging, Molinsky said, because of the heterogeneity in disabilities. She and her colleagues examined the literature on obesity, diabetes, arthritis, dementia, and overall trends in morbidity and longevity to get a sense of whether older adults in the future will be more or less likely to have disabilities. "In the end," she said, "we found that despite some recent gains in disability-free years, the trends in obesity and diabetes convinced us that there really was not much reason to decrease that rate so we held it constant for each age, race, and household type." The results of this analysis predict that by 2035 there will be 31 million households that are headed by someone age 65 or older and that have at least one individual with a disability. Broken down by disability, there will be 17 million households with a mobility disability, 12 million households with a disability that limits self-care, and 27 million households with a disability related to household activity (see Figure 3-4).

Despite the growing demand for accessible housing, the nation's stock of accessible housing is small, Molinsky said. She and her colleagues evaluated whether different classes of housing—single-family detached, single-family attached, small multifamily with fewer than five units, mid-size multifamily with 5 to 49 units, large multifamily with 50 or more units, and mobile homes—had the accessibility features of single-floor living, no-step entrances, and extra wide hallways and doors. They found

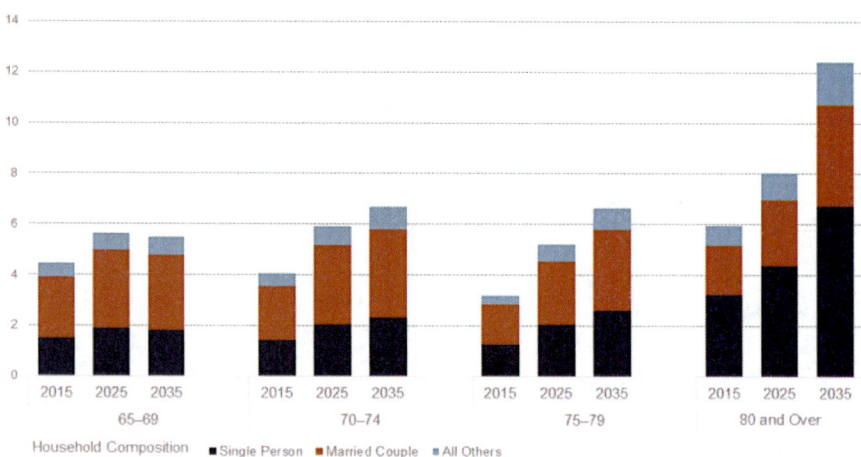

FIGURE 3-4 Projected households with disabilities by age group in millions.
SOURCES: Molinsky presentation, December 12, 2016. Joint Center for Housing Studies, *Projections & Implications for Housing a Growing Population: Older Households 2015-2035*, 2016, www.jchs.hardvard.edu. All rights reserved.

that only 4 percent of the nation's housing—public and private, although not including group quarters such as nursing homes and dormitories—has all three of these features (see Figure 3-5). Units in large multifamily buildings are most likely to have accessible features—in large part, she said, because they tend to be newer construction.

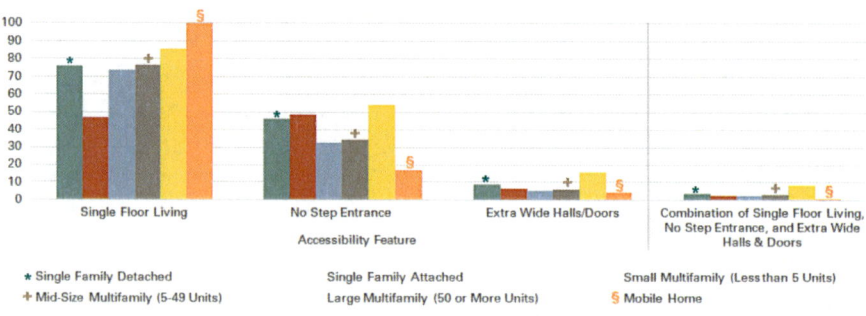

FIGURE 3-5 Share of housing units by structure type and accessibility features.
SOURCES: Molinsky presentation, December 12, 2016. Joint Center for Housing Studies, *Projections & Implications for Housing a Growing Population: Older Households 2015-2035*, 2016, www.jchs.hardvard.edu. All rights reserved.

Projections on Housing Affordability

Median income falls with increasing age for every demographic group after the mid 50s, which helps to explain why households headed by older adults are more likely to experience challenges with housing affordability, Molinsky said. People are said to experience a *housing cost burden* when they pay 30 percent or more of their gross household income on housing costs, including utilities, and a *severe housing cost burden* when housing costs, including utilities, account for more than half of the gross household income. For adults ages 65 and older, affordability depends strongly on tenure. For example, housing costs are burdensome for 17 percent of older adults who have paid off their mortgage, compared with 45 percent of owners who still have a mortgage when they retire and 55 percent of renters.

One repercussion of having unaffordable housing is that individuals will spend less on other necessities, Molinsky said. For example, lower-income older adults who are burdened by housing costs spend 67 percent less on transportation, 51 percent less on health care, and 37 percent less on food than lower-income adults who have affordable housing. In the future, she added, the number of individuals who face housing cost burdens will increase simply because of the expansion of the older population. "By 2035, we expect that nearly 11 million homeowners will face housing cost burdens, as will over 6 million renters," she said. "The number of the severely burdened is particularly alarming, projected to reach nearly 9 million among both of those groups by 2035." These projections, she explained, assume that today's income distributions will hold constant, but there are factors that could change those distributions, including trends in mortgage debt, income, and assets. One concerning trend, Molinsky said, is the rising share of older adults who enter retirement with mortgage debt, a rate that has been rising for the past 20 years. The median amount of mortgage debt at retirement has also been rising.

Household Income, Wealth, and Long-Term Care

Molinsky cited both hopeful and worrisome trends in household incomes for older adults. She said that more women are earning their own Social Security benefits and that both men and women are working beyond age 65. A caveat to this latter point, however, is that those who are working past age 65 tend to have higher incomes and less physically demanding jobs; low-income individuals are not necessarily working longer. One worrisome trend is that fewer people have traditional pension plans now than in the past, and the future of Social Security benefits is not certain. The Social Security Administration projects that there will be a decrease in the percentage of older adults who will be able to maintain their pre-retirement lifestyle into retirement, from 43 percent of older adults today down to 39 percent in the future.

Accumulated wealth can compensate for falling incomes, but wealth is not distributed equally across the population, Molinsky noted. For example, the median level of wealth among homeowners (when wealth includes the value of the home) is 42 times greater than the median level of wealth among renters. Even if home equity is excluded, the median home owner still has more than $100,000 in accumulated wealth, compared with $6,000 of accumulated wealth for the median renter. One result of this distribution of wealth, Molinsky said, is that while many older homeowners can afford long-term care for at least some period of time, most renters cannot (see Table 3-1). One study estimated that 69 percent of those who reach age 65 will need an average of 3 years of some type of long-term care at some point in their lives (Kemper et al., 2005). However, the average renter can only afford to pay for 2 months of long-term care. Of additional concern, Molinsky said, is that the ratio of family caregivers to older adults has been declining—family caregivers today supply the majority of care for older adults.

Molinsky also noted that while 15 million adults today earn less than 80 percent of their area's median income, which puts them in the low-income category, that number is projected to rise to 27 million by 2035. The number of older adults who earn less than half of their area's median income, which would make them eligible for federal rental subsidies, will grow from 4 million in 2016 to 7.6 million in 2035. "This is significant,"

TABLE 3-1 Most Older Owners Can Afford Long-Term Care While Most Older Renters Cannot

Care Category	Median Monthly Cost (Dollars)	Number of Months Before Median 65 and Over Household Spends Down Wealth		
		Renters	Owners, Including Home Equity	Owners, Excluding Home Equity
Homemaker Services	3,623	2	71	28
Home Health Aide	3,813	2	68	27
Adult Day Health Care	1,408	4	184	73
Assisted Living Facility	3,500	2	74	29
Nursing Home Care	6,448	1	40	16

SOURCES: Molinsky presentation, December 12, 2016. Adapted from Joint Center for Housing Studies, 2016.

Molinsky said, "because out of the current 4 million, we only serve a third of the eligible senior population." While this is slightly higher than the overall average of serving only one quarter of those in need, it still leaves a large number of older eligible adults without subsidies. "Even if we somehow figured out how to keep supporting one-third," she said, "we are still going to have 5 million older adults who are eligible for subsidies but who will not receive them and will have to find affordable housing in the private market."

Geographic Location of Housing

Based on data from the U.S. Census Bureau's American Community Survey and from the U.S. Department of Agriculture, Molinsky said that she and her colleagues have estimated that approximately three-quarters of older adults live outside of cities and that nearly half are aging in low-density locations—i.e., areas with less than one housing unit per acre. While urban living does not guarantee people will have access to services or that they will remain engaged in their communities, Molinsky said that people living in low-density areas generally face more challenges to accessing services and remaining engaged in the community. Most people want to remain in their homes for as long as possible, and if they do move, they usually do not move very far from home, Molinsky said. Aging in place can work, but not if individuals are isolated in their homes, she said.

When taken together, Molinsky said, all of the statistics she presented demonstrate that the nation must increase its supply of accessible housing. One possible policy solution, she said, would be to provide assistance to owners and landlords to pay for modifications that meet accessibility standards. Another policy solution would be to create incentives and regulations to ensure that new housing constructed in the future is built in compliance with standards for accessibility. Assistance for meeting accessibility standards could be in the form of grants, tax credits, or policies that would help owners safely tap into their home equity. While some of these options already exist, there is a need to scale them up to reach more older adults in more places across the United States, Molinsky said. There is also a need to increase housing options in the communities where people live. "So if half of the people are aging in these low-density suburbs and rural areas and they say they want to stay in those communities but their house may not be appropriate," Molinsky asked, "how about options for people to live in town centers, such as multifamily options or accessible dwelling units?" She said that she thinks there are many possibilities for creating more accessible housing, more affordable housing, and housing that older adults can more easily manage.

Molinsky also suggested that the nation build on promising pilot programs that support older adults with disabilities and health challenges in the home, including the 12 million people who have self-care disabilities. Molinsky's final recommendation was to consider how to support the 7.6 million low-income adults in their search for affordable housing so that they do not need to cut back on food, care, and support. "The sheer growth of the older population means there is much work to be done," she said.

DISCUSSION

Dara Baldwin of the National Disability Rights Network remarked that the discussions that she often hears portray disability and aging as something wrong or bad. She reminded the workshop participants that because of the Americans with Disabilities Act, individuals with disabilities have a civil right to live in the community. Sevak agreed, adding that researchers and policy makers have the obligation to develop policies that respect those civil rights. Molinsky added that the solutions that work for older adults should work for everyone, which is the principle behind universal design.

Phyllis Meadows from The Kresge Foundation asked the panelists if they had any insights into whether building development across the nation is displacing aging and vulnerable populations. Molinsky replied that the Joint Center for Housing Studies is evaluating how pressure from neighborhood changes affects vulnerable people. Sevak said that in many cities, developers are constructing accessible multiunit structures, but many of them are luxury units and therefore are not affordable for middle-income and low-income individuals. She said that it is important that the national conversation around housing include discussion of housing that is both affordable and accessible.

Daniela Koci of Loveland Center, Inc., asked if the panelists knew how to identify the best financing approaches to help support the housing needs of those who are living with disabilities or aging into disabilities. "To progress, we may need to go the extra step of asking the questions of would it be helpful to have certain types of tax credits, tax policies, or other types of approaches," she said. Molinsky replied that there are many piloted models available, including various kinds of loan and grant programs, but it can be challenging to bring these ideas to scale. She noted that Congress is currently considering a tax credit to help older adults and those with disabilities modify their homes. Another challenge is to identify the kinds of incentives that can be offered to landlords who own multifamily housing to convince them to improve accessibility by making modifications. Similarly, she said, there need to be policies in place to help homeowners safely use the equity in their homes—beyond offering reverse mortgages and home equity loans. Nonetheless, she said, there is no one-size-fits-all approach, and solutions will need to be tailored to individuals.

Koci also asked if there are any national efforts to provide better data regarding the income and needs of specific subpopulations of individuals with disabilities. For example, Koci said, someone with a developmental or intellectual disability would have different housing support needs than someone with diabetes. Sevak replied that some of the data in her presentation came from the Annual Disabilities Statistics Compendium,[2] which the University of New Hampshire releases each year. Those data break down disability into subgroups, but they do not include individuals with intellectual or development disabilities as a subgroup. The Social Security Administration does provide data with fine levels of detail, but those data are limited to individuals who receive federal benefits. Sevak also noted that whether a health condition becomes disabling can be a function of environmental factors. While nobody would question whether an individual born with Down syndrome has a disability, she said, someone with poorly managed type 2 diabetes is likely to suffer kidney disease, amputation, blindness, or one of many other disabling outcomes. "I think it is important, if we want to reduce rates of disability and increase rates of functioning, whether it is employment or other kinds of functioning, to think about health conditions that can lead to disability and intervene," Sevak said.

Margaret Campbell of Campbell & Associates noted that the surveys that are used to identify and characterize disabilities in the general population do not use the same categories as those that look at disability among older adults. The result, she said, are two different and inconsistent frameworks that need to be harmonized. In the same way, most surveys do not account for the fact that an individual can have multiple disabilities. "Just because a person has intellectual and developmental disabilities does not mean they do not have a hearing impairment or a vision impairment and vice versa," she said, adding that she thinks data exist that may be able to put various conditions together to look at different levels of need. The U.S. Department of Health and Human Services' Multiple Chronic Conditions Initiative[3] may be able to provide such data, she said, while also noting that identifying and specifying target populations in a sensitive and more accurate manner is important in terms of bridging the aging and disability fields and to forming partnerships.

---

[2] For more information, see http://www.disabilitycompendium.org (accessed February 15, 2017).

[3] For more information, see https://www.hhs.gov/ash/about-ash/multiple-chronic-conditions/index.html (accessed February 21, 2017).

Craig Ravesloot from the Research and Training Center on Disability in Rural Communities at the University of Montana asked about the availability and quality of data for assessing the needs for housing in rural America. Sevak said that basic data on the adequacy of the rural housing stock do exist, but more data are needed to understand how transportation serves that housing, how people in rural areas access the services they need, and how they engage with their communities. A great deal of attention is given to age-friendly communities in large metropolitan areas, such as New York City or Atlanta, she said, but one project that she will be working on will look at the frail elderly in rural areas who may not be able to get out into the community.

Robyn Stone of LeadingAge commented that the nation has paid very little attention to the demand for community housing options over the near term. "There is very little research on options in the community beyond some multifamily housing, a little bit of work on mobile home parks, a little bit on co-housing, but there is almost nothing in the literature around the future," said Stone, who called this a public health issue. Perhaps thinking of housing as a social determinant of health and a public health issue could lead to some different types of investments in housing for low-income older adults and individuals with disabilities, she suggested, adding that the data presented by Sevak and Molinsky demonstrate that there are going to be more lower-income older adults in the future who may not be able to remain in their homes. The homeless elderly population is the fastest growing segment of the homeless population in the United States, she said. Molinsky noted that in her hometown, housing for older adults is running into the same "not in my backyard" issues that create obstacles for affordable housing, so thinking about these types of housing as a means of addressing a public health issue might address the barrier of public acceptance.

4

Design Features of Accessible Housing for Older Adults and Individuals with Disabilities

The workshop's second panel featured three presentations on design features needed in accessible housing to support the health, well-being, and independence of older adults and individuals with disabilities. Bryce Ward, the director of health care research at the University of Montana's Bureau of Business and Economic Research, spoke about the connection between accessibility in the home and quality of life for older adults and those with disabilities. Corneil Montgomery, a senior program specialist in Habitat for Humanity International's Aging in Place program, and Patricia Tedesco, the coordinator of the Home Access Program at the Vermont Center for Independent Living, then discussed their organizations' efforts to increase the stock of affordable and accessible housing. An open discussion moderated by Elena Fazio from the Office of Performance and Evaluation at the Administration for Community Living followed the three presentations.

LIFE STARTS AT HOME: LINKING HOME ENVIRONMENT AND QUALITY OF LIFE FOR PEOPLE WITH DISABILITIES

Bryce Ward
Director, Health Care Research
Bureau of Business and Economic Research
University of Montana

Bryce Ward of the University of Montana began his presentation by talking about his grandfather, a man who had been a very socially engaged person when he was younger. As his grandfather aged, however, his mobility became more limited, and in the last several years of his life he rarely left his bedroom other than to go to a dialysis center. Nonetheless, Ward said his grandfather remained lucid and would happily engage in conversation in his room. Ward said that later, once he was an adult, he wondered why his grandfather could not come downstairs and socialize with the rest of his family if he was mobile enough to go to dialysis treatment. "How much of it was his health, and how much of it was other things, such as his environment?" Ward asked, noting that his grandparents lived in a multistory home that was not particularly accessible. "How much better would his life have been, how much bigger could we have expanded his choice set, if we had been successful at convincing them they should have moved to a different house or invested more in modifying the house?"

Ward said that stories like his grandfather's story are not uncommon. The Bureau of Labor Statistics' American Time Use Survey,[1] for example, showed that people in poor health spend 4 additional hours watching television each day. Ward said he wondered if improving the way older adults and individuals with disabilities interact with their home environment would increase their choice set and improve their quality of life. He and his colleagues are in the early stages of testing this hypothesis.

Data from the U.S. Census Bureau's American Housing Survey[2] provide some support for this hypothesis, Ward said. These data show that although households with at least one individual who uses some form of mobility equipment—a cane, crutch, wheelchair, or powered scooter—are more likely to have accessibility features in the home, many households that might benefit from these features still do not have them (see Table 4-1 and Figure 4-1). For example, 72 percent of households with a member who uses mobility equipment do not live on the ground floor of their buildings and do not have access to an elevator. Similarly, nearly two-

---

[1] For more information, see https://www.bls.gov/tus (accessed February 27, 2017).
[2] For more information, see https://www.census.gov/programs-surveys/ahs (accessed February 27, 2017).

TABLE 4-1 Features in U.S. Homes That Can Be Inaccessible Based on Whether a Household Member Uses Mobility Equipment

Feature	Percentage of Households Where Member Uses Mobility Equipment	Percentage of Households Where No Members Use Mobility Equipment
Stepped entrance	57.2%	60.9%
Upstairs with no elevator*	71.6%	81.7%
No grab bars in bathroom	62.3%	86.7%
No entry level bathroom**	18.5%	20.9%
No entry level bedroom**	32.4%	42.2%

NOTES: *Apartments not on the ground floor. **Multistory units.
SOURCE: Ward presentation, December 12, 2016.

thirds of individuals who use mobility equipment live in homes that do not have grab bars in the bathroom. These rates are high across all regions of the United States, Ward said (see Table 4-1), and they span urban and rural communities (see Figure 4-1). In rural areas, Ward said, some 91 percent of people who live in multistory apartment buildings and use

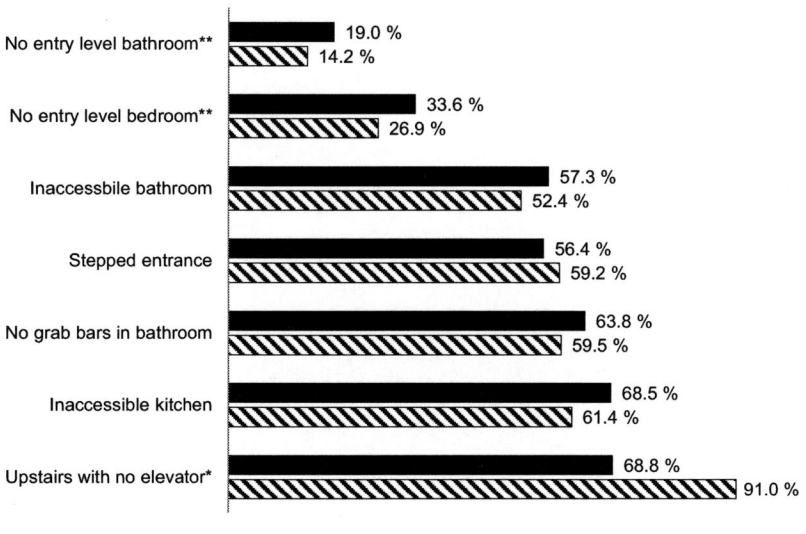

FIGURE 4-1 Rates of inaccessible housing among households with individuals using a mobility device in urban and non-urban settings.
NOTES: *Apartments not on the ground floor. **Multistory units.
SOURCE: Ward presentation, December 12, 2016.

mobility equipment do not have access to an elevator. "That makes it hard to engage in one's community," Ward said.

Ward and his colleagues are now conducting a study—called the Health and Home Survey—for which they surveyed households in three communities across the United States and asked survey participants about features of their home, how they feel while using those different features, and what percentage of maximal exertion is required for individuals in the home to engage in home activities, such as activities of daily living, entering and exiting the home, and preparing meals. For people with a mobility impairment, Ward and his colleagues found that these activities require two- to three-fold more exertion than for a person without a mobility impairment. In other words, Ward said, people with a mobility impairment have a harder time engaging in their houses. Moreover, the data also show that people who have a mobility impairment and could use an accessibility feature, such as a grab bar, exert far more energy engaging in the related activity, such as bathing, than do those who do not have a mobility impairment. "This suggests, but does not prove, that perhaps we can change this by changing our housing environments," Ward said.

Data from the U.S. Census Bureau's American Time Use Survey show that people without impairments spend about 33 percent more time—i.e., 38 more minutes—each day on household activities than do those with a mobility impairment. This includes activities such as gardening or food preparation and cleanup. Similarly, Ward said, people with impairments are less likely to report any time spent washing, dressing, or grooming, and when they do engage in these activities it takes them longer to accomplish them. These limitations matter, Ward said, because people who do not wash, dress, or groom themselves are substantially less likely to leave their home.

Ward said that his work has also shown that there is a 30 percent decline in the rate at which people say they participate in social and recreational activities for every standard deviation increase in how much time they report exerting themselves while bathing. While not conclusive yet, these data provide a link between living in houses that have many features that are not accessible and being less likely to get out into the world and engage in the social or recreational activities that are a determinant of quality of life. However, Ward said, getting people out of the house is not the only important issue, as there is also evidence suggesting that being able to do things at home such as housework makes people with mobility impairments happy and increases their sense of purpose.

Taken together, Ward said in conclusion, the data that he and his colleagues have collected suggest that it is important to enable people who have a disability to engage in more activity, whether it is engagement in activities in the home or in activities to get out of the home. "A potentially important part of allowing us to do all that is making your home more livable and more accessible because life does start at home," he said.

HABITAT FOR HUMANITY: HELPING PEOPLE AGE IN PLACE

Corneil Montgomery
Senior Program Specialist, Aging in Place
Habitat for Humanity International

Habitat for Humanity International's strategic vision is to have community impact by improving housing conditions, to have sector impact by partnering with other organizations to increase shelter access, and to have societal impact by inspiring action to end poverty in housing, said Corneil Montgomery of Habitat for Humanity International. Helping people age in place aligns with that strategy in three ways: it leverages shelter as a catalyst for community transformation; it increases the capacity to serve the most vulnerable; and it mobilizes volunteers as hearts, hands, and voices for the cause of adequate and affordable housing. Montgomery said that in 2015 approximately 13,000 volunteers ages 65 and older participated in the organization's projects in communities across the United States and that the organization prides itself in the way it uses older adults as volunteers.

Montgomery said that the national office of Habit for Humanity International has set a goal for itself to assess the housing needs of older adults, which means assessing not only the physical infrastructure of their homes but also considering the aspirations, preferences, and needs of the individuals living in those homes. In addition, as part of its Aging in Place program, the national office is encouraging local offices[3] to engage in larger community improvement projects, such as streetscaping, building community gardens, and helping local residents improve their financial literacy skills. He said, too, that some of the local offices are developing and testing new models of housing, such as townhomes and co-housing, from which the national office can learn and then share best practices throughout the organization.

The Aging in Place program has adopted a holistic approach to housing that requires partnering with community-based organizations to respond to a variety of needs in a community, such as health, transportation, mobility, and food access. The local office in Savannah, Georgia, for example, has a

---

[3] Montgomery explained that each of the 1,400 local Habitat for Humanity International offices in the United States is an independent, nonprofit organization. The national office can advise the local offices, but it cannot dictate what they do.

partnership with Meals on Wheels America[4] to study the effect that housing has on food security. This partnership also uses the volunteers of Meals on Wheels America to comprehensively assess the homes to which they deliver meals. Meals on Wheels America's assessment tools and questionnaires also enable Habitat for Humanity International to better understand the individual needs of families. Recently Habitat for Humanity International has also begun collecting data to determine the value of having universal design features in the homes it builds—i.e., no-step entry, wide doorways and hallways, easily accessible controls and switches, easy-to-use handles, and one-floor living—by measuring falls prevented, emergency room visits, quality of life, and cost savings to Medicaid and other parts of the health care system.

In fiscal year 2016, Montgomery said, local Habitat for Humanity International affiliates completed 6,094 repairs, which is a substantial increase from the 1,800 repairs completed in fiscal year 2015. Of the homes repaired in fiscal year 2016, 43 percent had an older adult living in the home, and 33 percent had an individual with a disability living in the home. Single women were head of household for half of the repaired homes, while single men were head of household for 14 percent of the repaired homes; 20 percent had at least one child under the age of 18. Montgomery said that when a local chapter works with families on repairs, it does not simply make the repairs and leave, but rather it partners with the families to learn about all of their needs and then coordinates with other organizations to address those needs that Habitat for Humanity International cannot address. The local chapter also conducts a pre- and post-repair assessment to find out if these families love where they live, if their health improves, and if the repairs prevent falls and improve other indicators of well-being.

Montgomery said that Habitat for Humanity International has several key accessibility improvements that it would like to install in all homes (see Box 4-1) but this is not possible, so the organization partners with families to both empower them and provide them with advice on best practices for home accessibility. The majority of Habitat for Humanity International homes have 80 percent of these features, which has been accomplished through partnership with families and local funding organizations.

Families that receive housing benefits from Habitat for Humanity International must participate in the building, rehabbing, or repairing of their homes. Those who cannot perform physical tasks can participate by engaging experts and the community. In order to fund these

---

[4] For more information, see http://www.mealsonwheelsamerica.org (accessed March 2, 2017).

> **BOX 4-1**
> **Habitat for Humanity International's Key Accessibility Features**
>
> - Single-story home
> - 42-inch wide hallways and stairways, with double handrails on stairways
> - Smooth, durable flooring
> - 36-inch wide doorways in the interior and exterior of the home
> - Zero-step entryways from the porch, the front door, and the garage
> - Zero-step entry into the shower
> - 60-inch wide circular area in the kitchen, laundry area, and bathroom
> - Electrical outlets that are 18 to 24 inches above the floor
> - Rocker-style light switches
> - Lever-style door and faucet handles
> - Accessible storage
>
> SOURCE: Montgomery presentation, December 12, 2016.

projects, Habitat for Humanity International provides its local affiliates with price repayment guidelines that are on a sliding scale. A typical payment for a repair will cost the family in the range of $50 to $100 per month for 36 months, although the local affiliates try to leverage as many local resources as possible to keep costs at a minimum for the families. Montgomery said that these projects use some local U.S. Department for Housing and Urban Development (HUD) funding. In addition, some affiliates leverage funding from Medicaid to make the modifications. The Greater Memphis, Tennessee, affiliate and Salt Lake Valley, Utah, affiliate are already leveraging pay-for-success funding and studying whether they might be able to use health care system community benefit funds to do repairs and improvements that can help people stay healthy and in their homes.

Looking ahead, in the short term Habitat for Humanity International plans to increase the number of home repairs it does by 5 percent, to enable greater mobility and visitability for 20 percent of the homes repaired, and to reduce the risk of falls in 20 percent of the homes repaired. Over the next 3 years, the national office will work with 10 local affiliates in a variety of communities to develop more evidence-based practices and to track, document, and analyze data to determine the cost–benefit ratio for specific housing and non-housing interventions in terms of fall prevention, food insecurity, isolation, and mobility. Over the long term,

Montgomery said, the goal is to affect the low-income older-adult generation so that more of them are able to stay and thrive in their homes and communities of choice.

THE VERMONT CENTER FOR INDEPENDENT LIVING: IMPROVING HOME ACCESS

Patricia Tedesco
Coordinator, Home Access Program
Vermont Center for Independent Living

The Vermont Center for Independent Living (VCIL), a statewide nonprofit organization, opened its doors in 1979. The Home Access program was added in 1983 to fund and build accessibility features for households where there is someone who has a permanent physical or mobility disability and where the household income is below 80 percent of the HUD median household income for the homeowner's county, said Patricia Tedesco of VCIL. Renters and families living in mobile homes can also apply for assistance, but only with explicit permission from the property owner.

After someone applies for assistance, the program sends a form to the applicant's physician to verify that the applicant has a disability and that the proposed home modifications would enable that person to live more independently or avoid moving into a nursing home. This also ensures a direct relationship between the home access program and an individual's health care provider, Tedesco said. Once program staff approve an application and put a project on the wait list, the applicant receives a "possible donor" letter that he or she can take to public service groups, faith-based organizations, or any other source of funding. The letter explains what the project is and how much the applicant needs to raise as "leverage funding"—$2,000 for a bathroom modification or $1,000 for a ramp, for example. The program also connects the applicant with a VCIL peer advocate counselor who can help the applicant look for leverage funding. At the same time, Tedesco said, she also begins "hunting for money." If the applicant has multiple sclerosis, for example, she contacts the National Multiple Sclerosis Society, which has funds set aside for these types of projects. Tedesco said that volunteer time counts toward leverage funding using a formula the program developed to calculate the value of the labor it would have to pay for in the absence of volunteer labor. The one exception to the leverage funding requirement is if the desired modification will enable an individual to leave a nursing home and return home. "If they cannot get into their home because they do not have a ramp or their bathroom is not accessible, and that is the only reason they are in the nursing home, no leverage funding is required and they move to the top of the list," Tedesco said.

Once the funds are in place, the program turns to one of five independent access consultants in Vermont who visit the applicant's home and develop a plan for the proposed modifications. The scope of work is then sent to three contractors whom VCIL has vetted and trained in an annual contractor seminar. "We work with a vulnerable population, so everybody [providing services] has to have a background check, references, and insurance," Tedesco said. Once she receives the three bids, usually within 4 weeks, she issues a contract, the project is built, and the access consultant performs a final inspection.

Projects that the home access program has completed include increasing the space beneath a bathroom sink to enable wheelchair access, converting a bathtub into a shower, installing fold-down grab bars, and installing ramps. In fiscal year 2016, Tedesco said the VCIL home access program received $545,000 of funding from the State of Vermont. The program completed 71 projects for 63 individuals with $181,245 of leverage funding and $28,995 of in-kind donations. The completed projects included 33 bathroom modifications, 35 ramps, 2 entrance modifications, and 1 platform lift. The average cost per project was $7,676. The 63 individuals served were an average of 60 years old, included 17 renters and 46 home owners, and had an average household median income of 39 percent of the HUD county median. Over the course of the year, the program received 98 applications, and 51 individuals were on the waiting list at the end of the fiscal year.

Lessons Learned

Tedesco said she has learned several lessons from her work on the home access program. Some lessons were easy to address, such as sending applicants a letter acknowledging that their application has been received, which reduces follow-up phone calls. Another lesson learned was to develop a landlord acknowledgment form, which not only provides details of any proposed modification to the home of a renter but also an invitation to the landlord to attend the initial inspection and to contribute to the project. The form notifies the landlord that he or she will be responsible for any future repairs to the improvements. In the case that an applicant owns their mobile home but does not own the land it sits on, the mobile home park owner needs to acknowledge that the program is building a ramp. Tedesco noted that the program is also moving away from installing permanent ramps to installing modular ramps that can be moved when an individual moves homes.

VCIL's primary funder has placed a cap of $15,000 on any single project, so Tedesco used to approve projects with budgets up to $15,000. However, when one project went over budget, she discovered she did not have a mechanism to pay for the additional cost. Tedesco has since capped projects at $13,500 to allow for change orders if they become necessary. Another financial lesson was that applicants must provide their portion of the project's funding before the project begins. The program has also partnered with Bath Fitter, which creates well-explained financial programs for bathroom retrofitting. She said that she is always seeking feedback from contractors, the access consultants, the people and organizations who provide leverage funding, nurses, and anyone else who can help ensure that she and her coworkers are not making decisions in a bubble but are working collaboratively. "I think that is what leads to some of our success," she said.

Tedesco concluded her presentation by noting that she and her colleagues are now hosting annual training on the Americans with Disabilities Act guidelines, VCIL's specifications, and disability awareness for contractors and funding partners. They are also building and supporting relationships with NeighborWorks America[5] and other partners that she said she hopes will streamline the funding process and produce better outcomes. Tedesco said she has testified at Vermont's House and Senate appropriations committee hearings to provide evidence of the importance of home modifications.

DISCUSSION

Winston Wong from Kaiser Permanente noted that some of the nation's more progressive health care organizations are looking at how to integrate social determinants of health into electronic health records. These organizations are also looking to take more accountability for the social aspects of keeping individuals as healthy as possible and optimizing the health care organizations' ability to care for their consumers' chronic conditions. In that context, he asked Montgomery if Habitat for Humanity International has a mechanism for sharing information with its clients' health care providers and health plans. Montgomery said that Habitat for Humanity International has learned from its Greater Memphis and Salt Lake Valley affiliates, which have strong connections to local hospital systems, about the value of such connections and is starting to have discussions with health care systems. One pilot project, for example, will deploy technology that will allow Habitat for Humanity International to coordinate information sharing with non-housing partners, including health care systems.

---

[5] For more information, see http://www.neighborworks.org (accessed March 2, 2017).

José Nuñez from the District of Columbia Department of Housing asked Ward and Montgomery if in their work they consider accessibility from the street to the home in addition to accessibility within the home and if there are federal or state grants to defray the costs of external improvements. Ward said that while the majority of the questions in the American Housing Survey and his program's Health and Home Survey are about accessibility inside the home, there are survey questions that cover external accessibility from the street to the home. For example, one question asks, "Considering all methods, can you get from the street to the house without going up at least one stair?" Montgomery said that Habitat for Humanity International's home assessment starts at the street curb, but added that the organization could do a better job of defining external accessibility more explicitly. He added that as Habitat for Humanity International moves from a one-house/one-family approach to a more coordinated approach with communities, the organization will be evaluating accessibility features of entire city blocks.

Robyn Stone of LeadingAge commented that the architects and other consultants she works with have found that lighting is important to accessibility, particularly for individuals with a visual impairment, and that the necessary modifications can be as simple as changing light bulbs. Montgomery replied that access lighting is a critical component of the modifications that Habitat for Humanity International recommends to its affiliates. The organization also recognizes lighting as being an easy to fix and sees it as something that the organization can empower families to do themselves. Tedesco said that her program is limited to making bathroom modifications, widening doorways, and building ramps, but that the program does address lighting that is inside the bathroom. Ward added that questions about lighting are included in the most recent version of the Health and Home Survey.

Emily Rosenoff from the U.S. Department of Health and Human Services asked Montgomery and Tedesco to talk about what gets people to finally ask for help from their programs. Tedesco replied that her program gets calls every day from people in a variety of circumstances. The challenge is that people call and expect to have a ramp installed in a month, not realizing that there is a 2-year waiting list unless they already have leverage funding available. Montgomery responded that most of the calls his program receives occur when people are in crisis, when repairs are needed to address a significant threat to health and safety and the family cannot afford those repairs, or when a doctor or occupational therapist has noted that repairs are needed before the individual can return home. Habitat for Humanity International is trying to be more proactive in raising awareness about its program, he added, so that people can begin planning for their needs in advance rather than waiting until there is a crisis.

Given that housing is a public health issue, Teresa Lee of the Alliance for Home Health Quality and Innovation asked if there could be a way to incentivize primary care providers to ask their patients during annual wellness visits about mobility and accessibility in the home, similar to how primary care providers address advanced illness planning and advance directives. Sarah Szanton from the Johns Hopkins School of Nursing said the Medicare annual wellness visit could serve as a means of being proactive about identifying potential accessibility issues. She said that she thinks this wellness visit should take place in the individual's home every few years and be reimbursed by Medicare. "Disability . . . is the relationship between the person, what they can do, and what their environment requires. If you are only talking to the person [and not seeing them in their home], you are not seeing that potential gap," she said. Ward replied that people should have conversations about accessibility not only with their health care provider but also with real estate agents and home builders. "There is all this information that we should start accumulating and start sharing," he said, "so that people are primed to start asking the question, not at crisis time, but when they are making a housing choice because, as we know, once they get into a house, they frequently do not want to leave it."

Miriam Kelty of the Washington Area Villages Exchange remarked that in Montgomery County, Maryland, homes with older adults experience fires and death caused by fires more than other homes, and she asked the panelists if their programs include installing smoke detectors. Tedesco said that the access consultant form the VCIL home access program uses includes questions specific to smoke detectors. In addition, the program installs a smoke and carbon monoxide detector in every home in which it does a bathroom modification or ramp installation. Montgomery said that Habitat for Humanity International takes the same approach as VCIL in its inspections and modifications. Kelty added that there are several young technology companies developing home safety equipment and that are looking for opportunities to demonstrate their technologies and thus they may be good partners. Ward said that his program is working on a study to evaluate these types of technology interventions and build an evidence base for programs such as Montgomery's and Tedesco's to use. Margaret Campbell of Campbell & Associates reported that IBM is doing a study in which it has partnered with postal carriers in Japan to report on the accessibility and livability issues they see in the homes of the older adults on their delivery routes. The carriers use iPads to report any issues they see. Campbell agreed with Ward and Kelty that data will be critical to determining the need, potential outcomes, and cost-effectiveness of potential technology interventions.

Jeffrey Henderson from the Black Hills Center for American Indian Health asked Ward if he had worked on housing accessibility with the Native American community in Montana. Ward replied that he has worked with that community on health issues, but not specifically on housing yet. Dara Baldwin of the National Disability Rights Network said that most government agencies do not work with Native American communities, though the Bureau of Indian Affairs is starting a program on housing, as is the National Council for American Indians (NCAI). The National Disability Rights Network is working with NCAI on accessibility issues through the Transportation Equity Caucus.[6]

---

[6] For more information, see http://www.equitycaucus.org (accessed February 27, 2017).

5

Models Connecting Affordable Housing and Services as a Platform for Health and Independence

The third panel featured presentations of specific successful and promising models of affordable and accessible housing in various regions across the United States. Peggy Bailey, the director of the Health Integration Project at the Center on Budget and Policy Priorities, provided an overview of the link between housing and health. The remaining panelists—Katina Washington, a program analyst in the Grants and New Funding Branch of the U.S. Department of Housing and Urban Development (HUD); Lisa Sloane, a senior policy advisor at the Technical Assistance Collaborative (TAC); Sarah Szanton, an associate professor of community and public health in the Johns Hopkins School of Nursing; Molly Dugan, the statewide director of the Support And Services at Home (SASH) program at Cathedral Square; and Rose Englert, a senior business leader of CareOregon's Community Health Innovation Program—then described their specific programs. Following the presentations, Craig Ravesloot, a research professor of psychology and the director of research and training at the Center on Disability in Rural Communities at the University of Montana, led a moderated discussion with the panelists and then opened up the discussion to the workshop participants.

SUPPORTIVE HOUSING TO IMPROVE HEALTH

Peggy Bailey
Director, Health Integration Project
Center on Budget and Policy Priorities

The Center on Budget and Policy Priorities,[1] Bailey explained, is a nonpartisan think tank based in Washington, DC, whose purpose is to ensure that low- and moderate-income families have access to the services they need. Bailey joined the center in January 2016 to lead its program Connecting the Dots: Bridging Systems for Better Health, which engages in cross-policy conversations that promote broad-based efforts to improve health at the population level. Her job, she explained, is to ask what work her colleagues are doing that is related to the federal Temporary Assistance for Needy Families program, how their efforts will help people with behavioral health needs, and what improvements they need from the behavioral health system.

The U.S. health system is focusing on housing and employment as social determinants of health, Bailey said, because there are successful models and opportunities for the integration of efforts. "From a housing standpoint, we know that lack of housing or poorer housing can dictate whether [individuals] are healthy or not," she said. "If someone is [living] on the street and has hypothermia, it is their lack of housing that is dictating their health, not just affecting their health in an indirect way." Therefore, it is relevant to talk about housing as a component of achieving the Triple Aim. "How can housing help reduce health system costs, improve health system quality, and improve access?" she asked, referring to the three goals of the Triple Aim. She said it is important to talk in terms that health care system leaders understand and to appreciate what those leaders are trying to achieve for the health care system. Meanwhile, those working in health care need to talk with their housing policy colleagues about what can be achieved through the housing system.

A number of small studies have shown that for high-need, high-cost populations, supportive housing can reduce costs while improving the utilization of the health care system, Bailey said. The best studies in this field have examined individuals living with HIV/AIDS (Aidala et al., 2007; Schwarcz et al., 2009) and have shown that supportive housing improves health status, lowers viral load levels, reduces the risk of death, and reduces risky behaviors. However, researchers have done little to show the impact that housing has on other chronic illnesses. More data linking better health outcomes with supportive housing could induce

---

[1] For more information, see http://www.cbpp.org (accessed March 8, 2017).

health systems to advocate for and invest in affordable housing as a means of meeting their mission and improving the measures by which they are judged.

Evidence does exist that shows that making housing improvements other than providing supportive housing has a positive effect on health. Research has shown that removing lead paint and managing the internal environment of a home to remove allergens and lung irritants can improve occupants' health (Sandel et al., 2010). Some managed care organizations have paid for air conditioning units, removing carpet, and taking other steps to decrease asthma attacks in children (Bryant-Stephens and Li, 2008; Sandel et al., 2004). More research is needed on the effects of overcrowding and housing instability on mental health, Bailey said. "There are things we intuitively think that housing can do, and we maybe have some evidence, but we do not have much really firm evidence . . . from a research standpoint," she said.

There is a range of types of housing models, Bailey said. There are housing services models that span single-site settings and provide onsite services as well as scattered-sites settings that aim to deliver services in homes distributed throughout the community. Housing specifically for older adults is an example of a single-site setting, and having services co-located with housing improves the ease of access, particularly for individuals with a disability. Single-site settings also make it easy for health systems to know where their clients are coming from. The main challenge for scattered-site settings lies in determining how to deliver the widest range of services possible in someone's home.

Designing programs to provide services can be challenging because individuals' need for services are not linear—they vary over time. "That makes thinking about profit margin and how much money [an organization is] going to earn difficult," Bailey said. Programs should be designed so that they provide services based on need, not on the amount of income a program needs to generate, she said. Another challenge is to match interventions to needs. "We need to stop making programs where we make people fail first so that we can figure out what they need," She said. "We need to do a better job of figuring out how to help people choose what is right for them and making sure that we can provide what they want."

Policy makers, Bailey said, also need to address workforce shortages, particularly given the effect that Medicaid reimbursement rates have on salaries and earnings. Furthermore, she said, service providers must do a better job communicating the value they bring to the health care system and developing cross-sector partnerships. "If we are thinking about the complex needs of people," she said, "we need to do better when it comes to partnerships." The same can be said for making the case for rental assistance. Bailey said that her organization has plans to conduct a big rental assistance initiative to increase rental assistance nationwide but is working on how to maintain the little money that is available. "It is going to take a collective effort of health care, housing, and everyone else who understands that affordable housing is important to health," she said. Other steps that need to be taken include making sure the low-income housing tax credit goes low enough from an affordability standpoint and protecting the Medicaid program, including strengthening Medicaid state plan amendments to affect the social determinants of health, she said.

Bailey recounted hearing Geoffrey Canada speak at a recent conference about the Success Academy Charter Schools[2] in Harlem; he challenged the audience to think about why it is necessary to prove that low-income people should have access to the same services as middle-income people. "How about we help figure out how to get those same kind of outcomes for low-income people?" Bailey said, adding that the Skid Row Housing Trust's Star Apartments[3] in Los Angeles, the National Church Residences program,[4] and the Camden Coalition of Healthcare Providers[5] are examples of programs that are successfully blending housing and health services for low-income individuals.

U.S. DEPARTMENT OF HOUSING AND URBAN DEVELOPMENT SECTION 811 PROJECT RENTAL ASSISTANCE PROGRAM[6]

Katina Washington
Program Analyst, Grants and New Funding Branch
U.S. Department of Housing and Urban Development

Lisa Sloane
Senior Policy Advisor
Technical Assistance Collaborative

Katina Washington of HUD began this joint presentation by explaining that HUD originally provided both capital and operating assistance to nonprofit organizations to construct or rehabilitate housing for individuals with disabilities. However, with declining appropriations and

---

[2] For more information, see http://www.successacademies.org (accessed March 15, 2017).

[3] For more information, see http://skidrow.org/buildings/star-apartments (accessed March 8, 2017).

[4] For more information, see http://www.nationalchurchresidences.org (accessed March 8, 2017).

[5] For more information, see https://www.camdenhealth.org (accessed March 8, 2017).

[6] For more information, see https://portal.hud.gov/hudportal/HUD?src=/program_offices/housing/mfh/grants/section811ptl (accessed March 8, 2017).

changes in the rental market, the capital investments were not adequate. The U.S. Congress passed the Frank Melville Supportive Housing Investment Act of 2010[7] to accelerate the development of affordable housing options for non-elderly individuals with disabilities, Washington said. The Melville Act added several new components under the Section 811 program, including the Section 811 Project Rental Assistance (PRA) Program, which provides rental assistance only to state-level housing agencies. The Melville Act encourages integration by requiring that no more than 25 percent of the housing can be designated for individuals with disabilities and that the units must be dispersed throughout a development. HUD works to ensure that tenants pay no more than 30 percent of their income for housing. In addition, services are made available to those who receive rental assistance, although participation is voluntary. The Melville Act also allows states to target specific populations with the greatest needs, such as people who are homeless or who are living in institutions.

When HUD developed the Section 811 PRA Program, Washington said, it wanted to ensure that services were in place for individuals with disabilities and that states' housing agencies were creating partnerships with states' health and human services agencies because state agencies often work in silos. In addition, the program incentivized states to develop innovative, replicable, and systemic strategies to provide housing with services for people with disabilities. HUD also wanted to determine an approach that would be faster than a brick-and-mortar process at increasing the number of accessible and affordable units available for individuals with disabilities and that would create more efficient and effective uses of housing and health care resources.

Today HUD places the onus of securing funding on the individual states, which have greater flexibility in putting together financing packages to develop this type of housing, Washington said. Many states use the low-income housing tax credit program, other HUD programs such as the HOME Investment Partnerships Program[8] and the Community Development Block Grant program, and funds from nonprofit foundations to construct housing units for people with disabilities. HUD then provides operating assistance in the form of rental assistance for these units. Washington said the agency encourages states to work with their local housing authorities to use existing public housing as additional housing for people with disabilities, or to use the Housing Choice Voucher program or other state-level affordable housing programs.

---

[7] Frank Melville Supportive Housing Investment Act of 2010, Public Law 111-374, 111th Cong. (January 4, 2011).

[8] For more information, see https://portal.hud.gov/hudportal/HUD?src=/program_offices/comm_planning/affordablehousing/programs/home (accessed March 15, 2017).

Congress funded new Section 811 PRA program units in fiscal years 2012, 2013, and 2014. The 2012 PRA demonstration awarded $89 million to 12 grantees which built some 3,000 units. In fiscal years 2013 and 2014 the program awarded $150 million to 24 grantees which built an estimated 4,500 units. More than 230 tenants currently live in units created by the fiscal year 2012 grants. Some 44 percent of these individuals came from institutions or were at risk of institutionalization, and 40 percent had been homeless or were at risk of becoming homeless. The units have stable tenure, Washington said, with about 8 percent of the tenants having exited the program for personal reasons.

Lisa Sloane of TAC then discussed how the state of Louisiana has implemented its Section 811 PRA program, which it established in 2012 when HUD awarded the Louisiana Housing Corporation (LHC) more than $8 million to create 200 PRA units. Louisiana's program focuses on central and northern Louisiana, including the cities of Shreveport, Monroe, and Alexandria, and was designed to build on the state's 3,000-unit permanent supportive housing program developed in the aftermath of Hurricane Katrina. Partnership between LHC and the Louisiana Department of Health (LDH) is a key component of the program, Sloane said. The collaboration was formalized through an interagency agreement that details roles and responsibilities for outreach, referrals, identifying target populations, and providing support services. An executive management council, consisting of the LDH deputy secretary, LDH program office assistant secretaries, the state Medicaid director, the LDH permanent supportive housing director, and the LHC housing authority administrator, meets at least once each quarter to assist with program implementation, management, and monitoring.

TAC has aided in Louisiana's development of its PRA program by encouraging the program to map how it plans to get an individual from housing application to tenancy (see Box 5-1). All of the partners have to work together and carry out their individual responsibilities to make the program work, Sloane said. "They sat down and put this together because if agencies and staff are not clear on their roles and responsibilities, you are not going to transition somebody from a nursing facility into an apartment in the time frame that unit has to be rented," she said. The program provides only 2 months of payments for vacancy, and transitioning an individual from a nursing home into a unit in that timeframe has been challenging.

There are program requirements and program preferences related to identifying the target population. To be eligible for the program, one must be between the ages of 18 and 62, have an income that is 30 percent or less of the area median income, have a disability as described in the statute, and be eligible for services at the time of admission. The Louisiana

> **BOX 5-1**
> **Steps in Leasing Process**
>
> 1. Affirmative marketing. Monthly call with Louisiana Department of Health and staff from various waiver offices and Continuum of Care. Louisiana Department of Health reaches out to the Continuum of Care and service providers.
> 2. Client submits application to Louisiana Department of Health Permanent Supportive Housing office.
> 3. Client is approved and added to appropriate waiting list. Client is sent a letter if ineligible. Client can request an appeal if determined ineligible, in which case Louisiana Department of Health responds within 10 days.
> 4. Property notifies Louisiana Department of Health that unit will soon be available.
> 5. Client is system selected and referred to provider. Provider has 48 hours to contact client and accept or deny referral. Louisiana Department of Health reviews whether there are already clients in system selected for this location and unit size. If not, Louisiana Department of Health does system selection. Select three clients per available unit.
>
> SOURCE: Washington and Sloane presentation, December 12, 2016.

program provides preference to individuals who are institutionalized or at risk of institutionalization, who are homeless or at risk of losing their housing, who have been affected by a hurricane, or who are youth aging out of foster care. When the program opened for applications, 585 people applied, of whom 188 had physical disabilities, 299 had serious mental illness, and 98 had an intellectual or developmental disability. The eligible properties have been identified—some units are currently filled and waiting for turnover to occur—and as of September 30, 2016, 43 tenants had been placed in 8 properties, with only 1 tenant having exited the program over the previous year. "Many of the tenants had difficult tenancy histories, including evictions," Sloane said, "but the Louisiana program has really managed to create a service and support system with a lot of stability." Services are largely funded by Medicaid, she added, and there is at least one certified service provider in each of the three cities. It took approximately 1 year to train the providers so that they could deliver services consistent with the model, Sloane said. "The partnership is working. The supports are there."

COMMUNITY AGING IN PLACE—
ADVANCING BETTER LIVING FOR ELDERS

Sarah Szanton
Associate Professor of Community and Public Health
Johns Hopkins School of Nursing

The Community Aging in Place—Advancing Better Living for Elders (CAPABLE)[9] program differs from many other care delivery models in that it is preventive—i.e., it aims to help people before they are in crisis—and it is framed as person-directed rather than patient-centered care, said Sarah Szanton of the Johns Hopkins School of Nursing. In addition, she said, CAPABLE is focused equally on health and housing, as opposed to being a housing program that might address health issues, or a health program that includes consideration of housing issues.

To provide an example of a typical individual participating in CAPABLE, Szanton told the story of Mrs. B. Mrs. B. has diabetes, hypertension, congestive heart failure, and arthritis—a set of chronic conditions common among older adults. She had lived in her home for many years and, as is true for many older adults, had fallen behind on home maintenance tasks to the point that her home had fallen into disrepair and large parts of it had become largely inaccessible to her. As a result, Mrs. B. was barely able to move around her home, spent most of her day in bed watching television, and ate poorly, all of which exacerbated her health conditions. Szanton said that those who have low incomes tend to have fewer resources to identify trustworthy people to help them fix things. She also said that she hears regularly from CAPABLE participants that in life, people receive training on how to be an adolescent and a young adult, but there is no training for aging. "You do not learn that you could have a three-dollar pull string on your ceiling fan and be able to turn the light on more easily," Szanton said. She also noted that people are more likely to be high spenders or high users of health care if they have a functional difficulty (see Figure 5-1). The health care system may focus on the chronic conditions that can cause functional difficulties, such as diabetes and congestive heart failure, Szanton said, but then it ignores the disability once the person is assessed as being "disabled." "The medical system does not assess for disability," she noted. "It does not assess whether you can take care of yourself, if you can stand long enough to cook or get your foot into the bathtub, but those are the things that will cause you to go into a nursing home."

---

[9] For more information, see http://nursing.jhu.edu/faculty_research/research/projects/capable (accessed March 8, 2017).

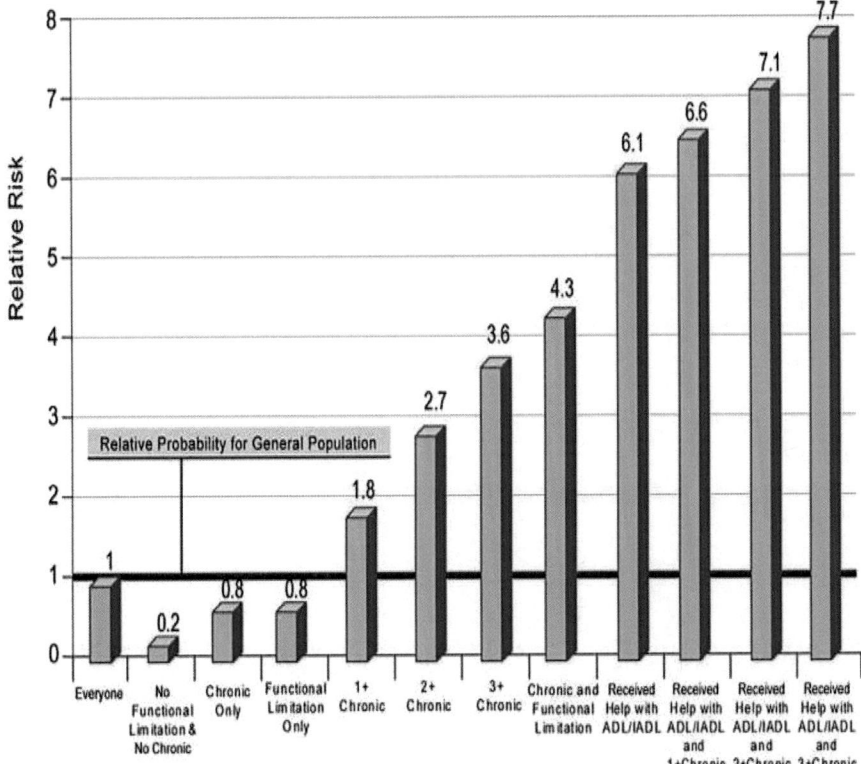

FIGURE 5-1 Relative risk of being in the top 5 percent of health care spenders for selected groups, 2006.
NOTES: ADL = activity of daily living; IADL = instrumental activity of daily living.
SOURCES: Szanton presentation, December 12, 2016. Data from Alecxih et al., 2010.

Szanton and her colleagues developed CAPABLE based on the success of the Advancing Better Living for Elders (ABLE) study conducted in Philadelphia (Gitlin et al., 2006; Jutkowitz et al., 2012). The ABLE study employed a combination of occupational and physical therapy sessions and home modifications to improve the ability of older adults to engage in instrumental activities of daily living (IADLs). Szanton said that, based on her experience as a home visit nurse practitioner, she and her team added nursing and home repair to the program. Each person's care program, which the individual directs, focuses on that individual's strengths, deficits, and goals with regard to IADLs and activities of daily living (ADLs).

The program is 4 months long and costs approximately $3,000—paid for with money from grants—which covers six visits from an occupational therapist, four visits from a nurse, and up to $1,300 in repairs and modifications by a licensed repair person.

During the first CAPABLE visit, the occupational therapist assesses the ADLs that the client prioritizes, such as bathing or dressing. In the second visit, the therapist watches how the client engages in those activities in his or her home in order to identify the modifications needed to enable the client to engage in those activities in a safe manner. Together, the occupational therapist and client create a work order. In Baltimore, the work order goes to Civic Works,[10] a nonprofit, job-training program that pairs an experienced contractor with an apprentice. The Civic Works team completes the repairs and modifications in the order in which the client listed them. Szanton said that all of the repair tasks must be strictly functional, not cosmetic. With the repairs and modifications underway, the nurse begins visiting the client at the beginning of the second month. Those visits focus on pain, depression, strength and balance, communication with the client's primary care provider, and medication. Over the remaining months of the program, the nurse makes three more visits and the occupational therapist works with the client four more times on the goals they identified.

Data from studies funded by AARP Foundation, the Centers for Medicare & Medicaid Innovation Center, The John A. Hartford Foundation, the National Institute on Aging, and the Robert Wood Johnson Foundation (Szanton et al., 2011, 2014a,b, 2015, 2016) showed that CAPABLE cut in half the number of ADL difficulties that participants experienced. Almost 75 percent of the participants, all of whom received Medicaid support, improved on measures of ADL limitations, and 65 percent improved on IADL limitations, such as managing medications, getting groceries into their home, and doing laundry (see Figure 5-2). Measures of depression also improved in more than half of the participants, with many individuals experiencing improvements in depression that were equivalent to taking an anti-depressant medication. "We are not giving them a pill," said Szanton. "We are giving them the ability to get out and get to church or to prepare a meal for a grandchild who comes over." The number of home hazards also fell in nearly 78 percent of the homes. Compared with a matched comparison group, individuals who participated in the CAPABLE studies required fewer stays in nursing homes and fewer hospitalizations. Szanton and her colleagues estimated that CAPABLE saves Medicaid an average of $10,000 per individual in medical costs. "That will be even more for Medicare because Medicare pays for hospitalization,"

---

[10] For more information, see https://civicworks.com (accessed March 15, 2017).

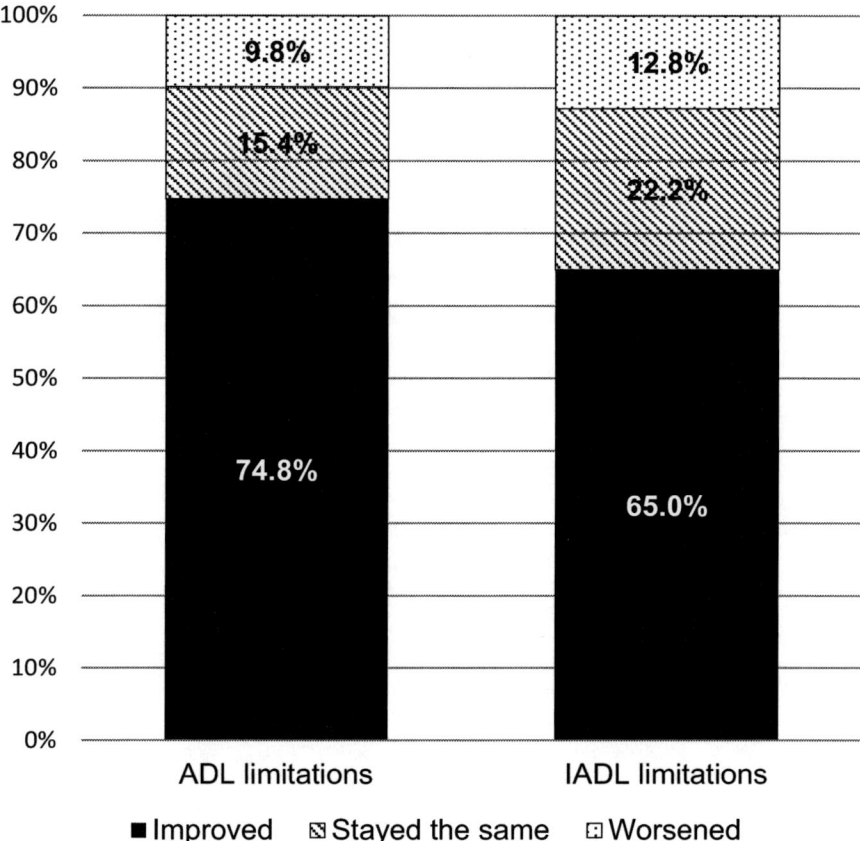

FIGURE 5-2 Changes from baseline to follow up in activities of daily living limitations and instrumental activities of daily living limitations among participants in the CAPABLE study.
NOTES: ADL = activity of daily living; IADL = instrumental activity of daily living.
SOURCES: Adapted from Szanton presentation, December 12, 2016. Data from Szanton et al., 2016.

she said.[11] Because CAPABLE costs approximately $3,000 per person, Szanton said, the cost savings is enough to pay for the program not only for the individual's participation but also for the participation of two additional people.

---

[11] Since the workshop, Ruiz et al., 2017, have published that CAPABLE saves Medicare more than $10,000 per person, per year.

Szanton concluded her presentation by describing the experience of the first CAPABLE client, Mrs. D. When Mrs. D. joined the program she was so disoriented that she sat on a commode chair all day, and the occupational therapist and nurse both thought she might be beyond the capacity of the CAPABLE program to help. On the first day with the occupational therapist, it took Mrs. D. 30 minutes to walk the 30 feet to her bathroom, which was why she sat on a commode chair all day. Mrs. D.'s husband, Szanton said, was so frail himself that all he could manage was to pull her out of bed in the morning, get her to the commode chair, and help her back to bed in the evening.

A review of Mrs. D.'s 26 different current prescriptions found that many of them were for pain and likely contributed to her disorientation. The nurse created a medication calendar and explained which medications were to be taken in the morning, afternoon, and evening. Soon after, Mrs. D. became more alert and able to talk with her care team about her goals, one of which was to be able to go downstairs in her home and wash her hair in her kitchen sink. She came up with the idea of placing plastic deck chairs along the way, about a foot apart, so that she could get up, walk to the next chair, and then sit for a while. After 1 month, she was able to travel the 30 feet to the bathroom, and by the end of the study she was able to walk the 30 feet in 7 minutes. Meanwhile, the repair crew installed a second railing on the staircase and added a light to illuminate the stairs, and once Mrs. D. was able to get down the stairs and wash her hair in the kitchen sink, she started going downstairs every day. "This was a woman who went from 'I do not know if we can do anything for her' to someone who could use her whole house," Szanton said, ending the story by reporting that a few months after Mrs. D. finished CAPABLE, her granddaughter called to say that the entire family was going to Atlantic City.

In summary, Szanton said, loss of function is costly. "It is what older adults care about, yet it is almost ignored in the medical system." Making simple changes to a home and providing short-term services can have a life-altering effect on the ability of older adults to retain function, remain independent, and live healthier lives, all at a relatively low cost.

SUPPORT AND SERVICES AT HOME

Molly Dugan
Statewide Director, Support And Services at Home
Cathedral Square

SASH uses affordable housing as a platform for realizing population health, said Molly Dugan of Cathedral Square. She commented that while the program's partners—home health agencies, mental health agencies,

Area Agencies on Aging, hospitals, and medical homes—had been working in the community for years to keep people healthy and at home, their efforts did not include participants from the nonprofit housing world. Noting that there are many symmetries between CAPABLE and SASH, she said that her organization would be implementing CAPABLE on the SASH platform beginning in January 2017.

Dugan and her collaborators started developing the SASH model in 2009 with the help of 54 residents in a senior housing site in Burlington, Vermont. Today, SASH serves 5,000 individuals with health care needs of every sort. Some participants are healthy and working full time, while others are extremely frail or have Alzheimer's disease. Once someone enrolls in SASH, he or she can remain in the program forever if desired. Some 80 percent of the participants receive Medicare benefits, 25 percent live in the community, 57 percent have a disability, and 27 percent are under age 65. Most of the participants live in traditional congregant affordable housing sites, though SASH does serve Medicare-eligible individuals in single-family homes, mobile homes, and private apartments. A Medicare demonstration grant funded the program through the end of 2016, and SASH will continue to receive funding through an all-payer waiver agreement between Vermont and the Centers for Medicare & Medicaid Services.

SASH is person-centered, Dugan said, and SASH staff members are embedded in the communities they serve. For example, the SASH care coordinator and the SASH wellness nurse have their offices in the affordable housing communities, enabling them to see participants regularly. The program is also oriented toward prevention and wellness, as opposed to crisis intervention, and relies on data-driven healthy living plans. Written memorandums of understanding formalize all partnerships with SASH. "Everybody knows what their roles and responsibilities are," Dugan said. "This is not lip service about partnership. It is real partnership. It is doing the hard work of explaining what we are all trying to do together and all agreeing to a certain set of rules." Partnership requires information sharing, and all SASH participants sign consent forms to allow for readily sharing information among the program partners.

Participants in SASH are assigned to a "panel" of clients served by one care coordinator, one wellness nurse, and a team of community providers that together build relationships that support being healthy at home, Dugan explained. Each month, staff members from the community providers meet with the housing-based staff, and SASH participants are always welcome to attend these meetings. "Basically, we are building this platform to provide population health with the housing organization as the host," Dugan said.

Cathedral Square, the nonprofit housing organization Dugan works for is 1 of 22 affordable housing organizations in Vermont that operate SASH out of 140 housing sites around the state. Some are in the most rural parts of the state, while others are in the most urban areas. Typically, panels have 70 to 100 participants, and panels in the rural parts of the state include individuals from multiple small housing properties. For example, SASH in Vergennes—which is in rural western Vermont—is based in the Armor Lane Senior Housing site, and 40 of the participants live there. The other 50 members of the panel live in the community, in either mobile homes or single-family homes near Armor Lane. The members in this panel average 76 years of age.

The SASH coordinator in Vergennes is Diana, who grew up in the town. When Diana enrolls someone in SASH, she conducts what Dugan called a person-centered interview in the person's home. This interview aims to learn about the individual, not just the individual's health needs. Dugan explained that staff members receive training in person-centered interview techniques. For example, during Diana's first interview with a new client, Larry, she learned that the 75-year-old was a sports fan, particularly of the Boston Red Sox, and that he liked to attend football games and other sporting events at the local high school. Larry also loved to cook and socialize. In a subsequent interview, Diana asked Larry about his goals, which were to be able to attend local sporting events again and to be able to cook his own meals again. Controlling his diabetes was not on his list of goals.

Next, the SASH wellness nurse conducted a uniform assessment with Larry during which she learned that he had diabetes with some associated nerve damage and that he had bacterial infections and generalized anxiety disorder. Larry also experienced frequent falls. With the information the nurse and Diana gathered, the SASH team created a healthy living plan for Larry, which included health coaching, regular check-ins with Diana to help with his generalized anxiety disorder, and attendance at a local diabetes class. The Area Agency on Aging designated a staff person to accompany him to sporting events and help him cook meals, and the home health agency made sure Larry was eligible to receive some personal care and housekeeping services. "It is truly a team effort," Dugan said.

The process Diana followed with Larry is the same process she uses for all 90 members of her panel. From the initial survey data, she learned that 54 percent of her panel had arthritis and 53 percent had hypertension. Dugan explained that Diana was charged with bringing in the appropriate partners and programming to meet those needs. She also operated from a community healthy living plan that focuses on exercise, walking groups, nutritious meals, mental health and social engagement, and other activities, such as offering a weekly blood pressure clinic, a foot care clinic, and a falls prevention class.

Dugan said that some 59 percent of SASH participants have documented advanced directives, compared to the national average of 26 percent, and the percentage of those who received vaccinations for shingles, pneumococcus, and seasonal influenza rose steadily over 18 months. The percentage of SASH participants who fell over the course of 12 months remained below 30 percent, which is lower than the World Health Organization's fall rate in the elderly. In addition, 77 percent of SASH participants had their hypertension under control, compared with the U.S. average among older adults of 30 percent. Finally, in 3 years of implementation, the growth in annual Medicare expenditures for SASH participants was lower by an estimated $1,536 per beneficiary.

PROMISING MODELS IN OREGON

Rose Englert
Senior Business Leader, Community Health Innovation Program
CareOregon

CareOregon is a nonprofit health plan and health services organization, Rose Englert said, and the Community Health Innovation Program is what she called CareOregon's social determinants team. CareOregon has some 230,000 low-income members, which is large by Oregon standards. It is a Medicaid plan and a dual special needs plan for Medicare. She said that the reason that CareOregon as a payer took an interest in housing was not because it has a dedicated community benefit fund—it does not—nor because it was required by law to do this work—it is not—but rather because CareOregon's leaders felt it was the right thing to do for their clients.

CareOregon has personnel specifically dedicated to housing, and the organization works with the Housing with Services consortium to provide onsite services. Staff members go into 11 low-income housing sites, and the program provides what Englert called light-touch health care and behavioral health care. When these care teams visit building sites, they assist residents in navigating health care services, regardless of insurance provider—or lack thereof—including helping veterans get care from the local Veterans Health Administration health care facilities. CareOregon staff conducts one-on-one follow-ups with its clients who received treatment either in the emergency department or as inpatients. These visits take place within 1 week and serve as a wellness check. They also help create and strengthen relationships with these individuals and get them more involved in their own health. "That is a huge part of the model," Englert said. "We cannot stress that enough."

CareOregon provides a range of services to its clients, Englert said. CareOregon has a transitions-of-care staff team that conducts a home visit after someone leaves the hospital rather than simply making contact via the phone call that Medicare requires. "It is amazing what you can learn with a home visit," Englert said. The organization also has a Go Mobile team that provides enrollment assistance for individuals in the community who need help finding care, figuring out whether they have insurance, or whether they are eligible for care as a veteran. The organization also has two housing case managers who work on internal and clinical referrals. Englert said that many of the people who receive the housing case management services have "burned out of the more traditional housing case management or even more intense community-based housing case management because of health issues." The housing case managers on her team have received training so they know how to work with the health care system to get the services these individuals require.

Englert said that her program has staff who are members of the Joint Office on Homelessness in Multnomah County and that her program is also working with the HUD Continuum of Care Program.[12] Together, they are trying to gain access to the region's Homeless Management Information System[13] to enable CareOregon members to get on the waiting list for HUD housing assistance. Englert said that CareOregon has made a capital investment in housing, working with five health systems to raise $21.5 million to address the shortage of affordable housing in Oregon. A large percentage of the $21.5 million fund is for construction of 3 buildings with 358 new units, including an onsite clinic with respite care beds for individuals who are homeless and trying to avoid hospitalization and recuperative care beds for individuals who have been hospitalized but do not have adequate housing to which they can return. Palliative care and hospice care will also be included. Englert said that the housing community should start thinking about palliative and hospice care when addressing the needs of aging adults. "We have to start thinking about hospice," she said, "because it is not just aging at home and staying at home—it is dying at home that is also really important." In fact, CareOregon is considering purchasing a hotel or an apartment space in southern Oregon, where there are no facilities that provide recuperative or respite care or

---

[12] For more information, see https://www.hudexchange.info/programs/coc (accessed March 16, 2017).

[13] The Homeless Management Information System is required by HUD and is used to collect client data on people experiencing homelessness or individuals who are at risk of homelessness.

that provide advanced illness care for older, low-income individuals. Her program is also working with care providers who make house calls as part of its palliative and hospice care initiative.

MODERATED DISCUSSION WITH PANEL SPEAKERS

Overcoming Program Barriers

Craig Ravesloot first asked the panelists to talk about some of the barriers their programs encountered while building their program or that have arisen now that their programs are operating. Dugan replied that with housing being the "new kid on the block," it became an early barrier for SASH. "We had to take a lot of time in meeting with [existing community] providers ahead of time and trying to get on the same page about what were the gaps that remained even though some of these organizations had been at this difficult work for 30 to 40 years." Through a series of "delicate conversations," she said she was able to convince the existing community-based providers that SASH could add value, such as with the trusting relationship that the SASH staff build with SASH participants. "We work where they live so they see us day in and day out," said Dugan. Even the home health care nurses who are in these buildings, she added, only see their clients after an acute episode or at most once or twice per week. "We were making the case that as housing staff on the ground, we would be able to fill in missing information, and that is how it has played out," she said.

Dugan noted that turnover in the leadership staffing of SASH's provider partners can be challenging because the new staff do not always appreciate the value of the housing-based staff. To address the turnover challenges, she invites executive leadership from all of the partner organizations to attend SASH team meetings. "They get to see how when you walk into these SASH team meetings it feels like you are walking into a meeting with people that work for the same organization. It is truly an integrated team," said Dugan. "I cannot say that we have completely overcome that challenge, but we know what we need to do to keep confronting it," said Dugan. In the same vein, Szanton said that she is frequently told, "We already do that," when talking about the services CAPABLE provides. In fact, Szanton said, such programs rarely take their lead from what their clients desire, nor do they supply their clients with the skilled labor to help them make changes.

Englert said a big challenge for her program has been addressing privacy barriers arising from the Fair Housing Act, the Health Insurance Portability and Accountability Act (HIPAA), and other policies. CareOregon will not partner with housing organizations unless they agree to put a release of information related to HIPAA and the Fair Housing Act in the tenancy packets, she said. She said that in her opinion, advocates should have a conversation with regulators about creating waivers for specific projects or rethinking how privacy laws interact with one another so that they do not create silos that prevent the kind of integration needed to address housing- and health-related problems. Szanton added that when her program thought about writing letters to their clients' primary care providers informing them about their patients' goals and what they have achieved, she quickly realized that they had to get people to first sign privacy waivers.

Bailey noted that while there are ways to overcome the privacy barrier, this work would be easier if policies related to information exchange were harmonized and widely disseminated. At the same time, she said that some of the larger payers have lengthy contracts they expect their housing partners to sign, assuming that they have lawyers on hand to help them navigate the legalese. Perhaps, she said, payers could be more flexible in thinking about what these contracts really need to include.

In terms of payment Englert said a barrier arises when programs restrict how funds can be spent. Oregon is fortunate, she said, because its Medicaid waiver enables organizations like CareOregon to spend money on what it calls flex or health-related services, although some policy changes would make it easier to get more payment models involved.

Sloane said she and her colleagues in Louisiana have found that staff retention makes a big difference in the long-term success of their programs, in large part because the programs are very complicated and have many moving parts. In time, she said, the infrastructure can become strong enough to continue on when key personnel turn over, but staff stability at the beginning of a program is important. Englert noted that strong workflows contribute to staff retention and continuity. "If you are basing your program on an individual, it is not a replicable model," she said.

Szanton referred back to a comment Erika Poethig made in her keynote presentation about the "wrong pocket problem," referring to the inability to capture the savings in health care expenditures that result from an expenditure on housing. "What we have tried to do is measure the outcome of all of the pockets," she said. Her program, for example, calculated what CAPABLE saves in terms of Medicaid spending. She then approached Medicaid and asked what portion of the savings it could spend on services that would add value to the system. She noted that the often-used approach of calculating quality-adjusted life years[14] is not

---

[14] Quality-adjusted life years (QALYs) is a measure often used to show potential savings from a social program or a health improvement intervention. A QALY value of 0 is equivalent to being dead, while a QALY value of 1 is equivalent to being in perfect health.

concrete enough for Congress. "If we can say we are saving you dollars or saving nursing turnover . . . I think that is where we can make a difference," said Szanton.

Factors Contributing to Housing Program Success

Ravesloot then asked the panelists to comment on the factors that contributed to the success of their programs. Englert said her program was lucky to have strong executive leadership that believed in and funded the program. "You have to have those organizations that have the guts to go out there and do this and try new things and evaluate," she said. Similarly, Dugan said that she made a point of identifying the executive directors in Vermont's affordable housing network who were willing to take a risk and go into health care, which was new for many of them. Vermont has six designated regional housing organizations that have the leadership to disseminate the SASH model statewide, and this core group has been implementing it since 2011. "If you have champions, you are better off because then you have that sustainability," said Dugan. She added that what brought this core group together was identifying the common challenge of housing an aging population and the common goal of not merely putting a roof over someone's head but also providing them with services to enable them to stay in their homes.

CAPABLE's success, said Szanton, comes from focusing on housing and the individual, and homing in on each individual's goals. In terms of demonstrating impact, the key has been to measure a wide range of outcomes, including stories and quotes from those benefiting from the program. For the Section 811 PRA program, Washington said it was important to have external technical assistance on a one-on-one basis, which HUD itself cannot provide. "In HUD, we do not have the knowledge or experience to give grantees personal attention, but with the technical assistance [providers] we have been able to do monthly peer-to-peer calls with grantees to talk about policy updates and give them the opportunity to discuss their issues and their successes," said Washington. The technical assistance mechanism, she added, enables the grantees to create working groups and focus groups without having to involve HUD.

The Section 811 PRA program has also benefited from looking at a broader range of outcomes using reporting that goes much deeper than it used to. Instead of just counting the numbers of people living in subsidized units, HUD is looking at where the residents are coming from and how they got into these units, for example. Washington noted that the program is now being evaluated and she said her hope is that this evaluation will highlight more successes and identify cost savings that come with moving from a brick-and-mortar approach to a more holistic approach.

Making a Business Case for Housing Programs

Ravesloot next asked the panelists how they make a business case for their programs. Englert replied that it is necessary to first articulate goals clearly at the very beginning of the program and that such goals need to go beyond saving dollars or reducing hospitalizations. She said she works in the context of goals that are SMART: specific, measurable, actionable, replicable, and time-bound (Doran, 1981). She also said when she tries making a business case based solely on overall cost savings, that is when she sees the different organizations bumping heads.

Bailey said the business case will differ depending on who one is trying to engage in a hospital system. "Telling them you are going to save money is not going to work because they are not in the business of saving money. They are in the business of making money," she said. Instead, a hospital administrator will be interested if the business case involves reducing the number of people who use the emergency department for routine medical care and who take time away from those who truly need emergency services and whose insurance will pay for those services. She suggested that when engaging a managed care organization, first tell the organization what a program does and ask them what parts of that program interest them and would help them meet their goals. When speaking to the government, a straight dollars and cents argument is appropriate because government policymakers want to know how much money an investment will save or how it will enable funds to be used more efficiently. Bailey recommended talking to peers in other states to find out what measures they are using and to ask them to share their outcomes data.

DISCUSSION

Teresa Lee asked if the cost-effectiveness of these programs changes when funding from foundations is added to funding from the federal government. Englert replied that CareOregon used federal government funds to prove the care model and now foundations are stepping in to fund dissemination in other places. For example, the Visiting Nurse Association of Colorado has received a grant to deploy CAPABLE as a means of reducing hospitalization rates for Kaiser Permanente clients in Colorado. Four cities in Michigan have received funds from the Hillman Foundation and Michigan Medicaid to deploy CAPABLE in those cities. Michigan Medicaid has pledged to make CAPABLE available to all 15,000 Medicaid beneficiaries in the state if the 4 cities can show that the program delays nursing home admissions by an average of 1 day. A workshop participant commented that tracking dollars is important but expressed fears that the Triple Aim combined with elements of the Patient Protection and Affordable Care Act have led everyone to believe that it is possible to deliver better quality care for less money and give money back to the system. This participant said that living in a community is a civil rights issue and an important thing to do regardless of whether it saves money or generates money.

Winston Wong of Kaiser Permanente asked the panelists if there is a way to empower the people whom these programs affect so that they can advocate for themselves, rather than only having service providers tout the benefits of these programs. "I think this is an equity question with regard to empowering the individuals in communities to say this is not a question of trying to convince someone but about seizing their right for integration into society," Wong said. Bailey said that there are examples of this and that it is an idea that she supports. "This is exactly what we want to do—to explain what happens to care for low- and moderate-income people if we are not thinking about this holistically," Bailey said. However, she added, her worry is that populations will be pitted against each other, given the limited funds that are available and the fact that states will have to make choices about who will and who will not receive coverage. She said that organizations such as Families USA[15] have been successful in helping organizations tell the stories of the people they serve. She also said that her organization has just announced it was a founding member of the Protect Our Care Coalition to protect Medicare and Medicaid and that part of that effort will involve telling stories of what providers are doing.

In the future, Bailey said, it will be important to hold officials accountable for the effects that their policy decisions have on social justice and social equity and to not be shy about doing so. "We have to call attention to that and explain why it is not acceptable," Bailey said. Dugan noted that the 5,000 Vermont residents who are enrolled in SASH have been very vocal about how the program has changed their lives and have made a difference in ensuring that the program has continued to receive funding. She added that it is important for those who benefit from these programs to feel invested in them. "I think that is where your compelling story for continuation is going to come from," Dugan said. Ravesloot said that the disability community has been a leader in this approach for 30 to 40 years. "I think that joining forces and breaking down silos creates a larger voice," he said. Englert suggested everyone read an article in *The New Yorker* by Atul Gawande that makes the case that the community needs to refocus its attention to make sure that the general population, and not just the people who attend workshops such as this one, understand why social

---

[15] For more information, see http://familiesusa.org (accessed March 17, 2017).

services are important, why health care is important, and why access to care is a basic human right (Gawande, 2016).

Dara Baldwin said her organization is always working to ensure that the different beneficiaries of non-defense discretionary funding are not making deals to cannibalize each other's programs. "I have to say that not all social justice groups are following this, and it is a shame," Baldwin said.

Bailey remarked that the Center on Budget and Policy Priorities has been known for its long papers filled with charts and graphs. Now, however, the center is turning its focus to conveying its message to the public and using its technical expertise to translate its message so that it speaks to everyone with regard to various races, geographic diversity, and religious diversity. Her concern is that every time she talks about the working class, people now assume she is talking about only white Caucasians, Christians, or heterosexual individuals. "How do we have a more holistic conversation without putting everyone into groups that are now targeted against each other?" she asked. "It is not easy, but it is something I can honestly say we are committed to. It may be bumpy at first, but it is something that has to happen in order . . . [to explain] what social programs do across the country."

6

Reactors Panel on Policy Implications and Research Needs

The final panel session was devoted to a conversation about the future of affordable and accessible housing, next steps, policy implications, and research needs. Emily Rosenoff, the manager of and a program analyst with the U.S. Department of Health and Human Services (HHS) Office of the Assistant Secretary for Planning and Evaluation in the Office of Disability, Aging and Long-Term Care Policy, moderated the panel, which included the following reactors: Dara Baldwin, a senior public policy analyst for the National Disability Rights Network (NDRN); Anand Parekh, the chief medical advisor at the Bipartisan Policy Center; Robyn Stone, the senior vice president for research at LeadingAge; and Uchenna S. Uchendu, the chief officer for health equity at the U.S. Department of Veterans Affairs (VA). Each reactor had 5 minutes to share his or her thoughts, and the discussion was then opened to the workshop participants.

REACTOR COMMENTS
Dara Baldwin
Senior Public Policy Analyst
National Disability Rights Network

Dara Baldwin of NDRN began by explaining that NDRN is the national organization for the Protection & Advocacy Network (P&A Network), which provides legal representation and advocacy for people with disabilities. The network is funded by the Administration for Community Living and has 57 programs, one for each state, the District of Columbia, and every U.S. territory, along with one that specifically serves Native Americans. Housing and employment are two major areas of focus for the P&A Network and NDRN. Baldwin noted that NDRN belongs to the Transportation Equity Caucus[1] and works with the National Council on Independent Living[2] in efforts related to housing, particularly issues centered on the concept of visitability.[3]

Baldwin said that veterans' groups are potential partners in housing efforts but were not mentioned during the workshop. While the perception is that these organizations only work with veterans, Baldwin said she has seen examples around the country where these groups help individuals who are not veterans but do have a disability.

Baldwin also noted several areas of research that were not addressed during the workshop:

1. Disaggregating data on intersectionalities.[4]
2. Reentry for those who were formerly incarcerated, many of whom are older when they are released, experience the onset of disabilities, and are also searching for housing. NDRN is working on a bill with the Reentry and Housing Coalition[5] related to reentry, but it needs data to support the proposed policies and to show that the programs that exist really do help formerly incarcerated individuals. The issue of housing is often lost in the criminal justice reform conversations. Baldwin noted that Medicaid expansion under the Patient Protection and Affordable Care Act made formerly incarcerated individuals eligible for Medicaid benefits. "We need the data on how Medicaid services helped those people who were formerly incarcerated," she said.
3. Accessibility of playgrounds and recreation centers.
4. Access to public housing and the public pathways that those with disabilities have to traverse to get from the street to their homes.

---

[1] For more information, see http://www.equitycaucus.org (accessed March 15, 2017).

[2] For more information, see http://www.ncil.org (accessed March 15, 2017).

[3] Visitability is a movement to build houses that include a zero-step entrance, doors with 32 inches of clear passage space, and a bathroom on the main floor. For more information, see www.visitability.org (accessed March 15, 2017).

[4] Intersectionality, a concept developed by Columbia University law professor Kimberlé Williams Crenshaw and originally articulated on behalf of African American women, is a term used to refer to overlapping social identities (e.g., race, gender, sexual orientation, religion, age, disability, etc.) and related systems of discrimination and oppression.

[5] For more information, see http://www.reentryandhousing.org (accessed March 15, 2017).

Anand Parekh
Chief Medical Advisor
Bipartisan Policy Center

Accessibility is critical, said Anand Parekh of the Bipartisan Policy Center, if for no other reason than that it helps address the public health challenge of falls in the home. One in four older adults fall every year, accounting for 2.8 million emergency department visits, 800,000 hospitalizations, and $31 billion spent annually in Medicare expenditures, he said. Parekh, who is himself a physician, said he was hopeful that the two cabinet nominees for the U.S. Department of Housing and Urban Development (HUD) and HHS, who are both physicians, might be able to find common ground in their work to accelerate integration of housing and health. He suggested that state and local officials raise their visibility on health and housing integration to help drive home the message that there are opportunities for these two federal agencies to make a significant impact in the lives of older adults and individuals with disabilities.

Parekh noted that the Bipartisan Policy Center organized a senior health and housing taskforce in 2015 headed by two former HUD secretaries, Henry Cisneros and Mel Martinez, and two former members of Congress, Allyson Schwartz and Vin Weber. In May 2016 the task force released its report *Healthy Aging Begins at Home* (Senior Health and Housing Task Force, 2016), which contains 30 recommendations to bring the housing and health care sectors closer together in order to improve health outcomes and reduce preventable health care costs for American older adults. The task force directed its recommendations to the executive and legislative branches of the federal government with the goal of expanding the supply of affordable housing, facilitating home and community-based modifications, and accelerating health care and housing integration in order to improve outcomes and reduce cost. Furthermore, the task force report recommended that federal efforts focus on the 1.3 million Medicare and Medicaid dual-eligible beneficiaries who live in publicly assisted housing and go beyond what HUD is doing with their supportive services demonstration. The idea would be to invite managed care organizations, accountable care organizations, and other providers to be accountable for the total cost of care and outcomes for this population. "To achieve these outcomes," Parekh said, "they would need to partner with housing providers and provide these evidence-based models you have heard about today [at the workshop], whether it is SASH [Support And Services at Home], CAPABLE [Community Aging in Place—Advancing Better Living for Elders], or other models. They would receive advanced payments, and if there are cost reductions they would share savings with housing providers and with Medicare and Medicaid."

Another recommendation in the task force's report was for states to take more advantage of Medicaid coverage for housing-related activities and services for older adult beneficiaries and those with disabilities. "We do not really have a clear picture of how states are utilizing these permissions, and I think there is a research opportunity there that can inform both federal as well as state policy," Parekh said. Rosenoff added that the Centers for Medicare & Medicaid Services (CMS) has been running an innovation accelerator program with a number of federal partners. This program has provided intensive technical assistance to eight states and less intensive technical assistance to some 30 other states to help them focus and strengthen their housing and health partnerships. As a result of this effort, Rosenoff said, approximately 10 states have applied or will soon apply for Medicaid waivers to provide supportive housing services as part of Medicaid benefits.

For Congress, the task force called for expansion of the low-income housing tax credit. One concern is that corporate tax reform could make this credit less valuable for private developers. Parekh noted that over the 30 years that this tax credit has been available, it has enabled construction of nearly 3 million low-income housing units at a cost of $100 billion. Another proposal calls for the Administration for Community Living to better coordinate the nine different federal programs run by five executive branch departments that help older adults modify their homes. "These programs are doing good work, but are not very well connected, nor are they coordinated," Parekh said. This proposal also called for using the aging network and the Aging and Disability Resource Centers to disseminate information to older adults about the availability of these programs. Parekh added that a number of U.S. senators have introduced a bipartisan bill calling for these same steps. Parekh concluded by saying that there will be opportunities for Congress to help low-income Americans by integrating health and housing. "As much as I think we do need more efforts at the local and state level, we remain optimistic that there might be movement at the federal level in 2017 and beyond," he said.

Robyn Stone
Senior Vice President for Research
LeadingAge

Robyn Stone of LeadingAge said that many countries see housing as a right, not like the lottery it seems to be in the United States. In Singapore, for example, housing was the bedrock of that country's development in the 1950s, and housing is seen as essential for well-being. This idea is relevant to the United States from a policy perspective because so many of the federal dollars that go toward housing are discretionary, which makes future federally funded support for housing for low-income older adults and younger adults with disabilities uncertain. She also reflected on housing becoming a "social determinant of health buzzword." "Coming out of public health as I do," she said, "we always knew that housing was an essential determinant of health." However, she said that in her opinion housing is much broader than only a social determinant of health. It is an often unrecognized public health issue for low-income older adults, adults with disabilities, and other vulnerable populations, which is different than talking about it just as a social determinant of health. Housing is shelter, which is a social determinant, she said, but it is also quality and accessibility and a platform for the effective delivery of services, supports, and prevention. This workshop, she said, brought together all of these aspects of housing and focused not just on moving toward the goal of the Triple Aim of health care but also on enabling people to live as independently as possible in the community.

Stone applauded the recent report from Harvard's Joint Center for Housing Studies (Joint Center for Housing Studies, 2016) and the workshop discussion about meeting the housing needs of older adults and younger people with disabilities. She said that they clearly highlighted the fact that there is a large low- and modest-income population which is growing, so the current stock of housing available is unacceptable and untenable for the future. "How are we going to solve that?" Stone asked, while also agreeing with Parekh about possible cooperation between HHS and HUD when they are both headed by physicians.[6] She noted, however, that "we are going to have to continue to look a lot at public–private partnerships and a more significant role for the private sector."

Together with her colleague, Stone is conducting a study on the feasibility of using social impact bonds[7] to boost health and housing partnerships for low-income elderly adults in Los Angeles. Most of the use of social impact bonds has been for individuals who are homeless or formerly incarcerated, which she said are populations at such risk that almost any investment will generate savings. "That is where private investors want to put their money," she said. "For the populations we are talking about today [at the workshop], it is much more nuanced in terms of where you get cost savings and how those dollars would ultimately go back to a private investor."

---

[6] The current secretaries of HHS and HUD are physicians Tom Price and Ben Carson, respectively.
[7] Social impact bonds are a form of funding whereby private investors provide the upfront capital for a project and the investments are repaid only when and if the targeted outcomes for the project are achieved.

Where it may be possible to influence the new administration is by framing this problem as part of the infrastructure needs of the country, and not just in metropolitan areas but in rural communities too, Stone said. "Perhaps thinking about this in terms of infrastructure will be a way to keep [housing] on the [national] agenda and also think about ways in which we can get money into these models and these programs other than through the health care route," she said. Stone concluded her comments by noting she does not believe that there is enough of an evidence base to persuade the government to invest in these housing programs. Her team's evaluation of SASH, which has continued through three cycles to find a slowing of the growth curve in Medicare spending, is the kind of evidence that health plans want to see, she said. Similarly, she said, there are promising signs with CAPABLE and some of the other models discussed during the workshop, but there is still a need for more evidence. "I think the potential is there, but the evidence base needs to be built," she said. Stone also expressed optimism that the move to value-based payments in health care will not go away, given that it has bipartisan support, and that the housing and health model will be shown to provide some of the best opportunities for value-based payments to have a positive impact on health outcomes and on improving delivery of services to high risk older adults and younger adults with disabilities.

Uchenna S. Uchendu
Chief Officer for Health Equity
U.S. Department of Veterans Affairs

Noting that she agreed with most of the comments of the three reactors who spoke before her, Uchenna S. Uchendu of the VA reiterated Baldwin's recommendation to reach out to veterans organizations and to begin asking the individuals who participate in the programs discussed at the workshop if they or their family members have served in the military. This is important, she said, because while there are many services that the VA can provide, it is only doing so for approximately 9 million of the estimated 22 million living veterans. "We cannot afford not to include all of them," she said. "We have a duty to honor Abraham Lincoln's promise— 'To care for him who shall have borne the battle, and for his widow and his orphan.' Today, we say 'for their families and their survivors' because women now serve in the military."

Uchendu said that part of her job in championing health equity and the elimination of health disparities for all veterans, but especially the most vulnerable, is to be the liaison to other agencies that are also working toward health equity. She noted that because less than 2 percent of the U.S. population serves in the military, veterans are a minority by numbers. Additionally, veterans' unique military experiences and exposures in different military periods or eras[8] add another layer of vulnerability. A combination of these factors increases the likelihood of health disparities for veterans.

The VA is a good model of where health care and social determinants such as housing intersect, Uchendu said. Veterans' benefits administered by the VA include education through the G.I. Bill, housing via VA loan guarantees, and housing the homeless through its partnership with HUD and other stakeholders. In addition, the VA will provide home modifications for veterans whose service has made it impossible or difficult for them to function in their homes. Because the VA brings income, housing, education, and health under one umbrella, it has data to support the effectiveness of programs such as homeless patient-aligned care teams and other person-centered programs that work at the intersection of health and various social determinants of health.

Rosenoff remarked that the efforts of the VA, HUD, and her department at HHS to address veteran homelessness are having great success. Since 2010, when the focus on homeless veterans began and Congress provided the necessary resources, homelessness among veterans has been cut by 50 percent, and three states—Connecticut, Delaware, and Virginia—have ended veteran homelessness, as have 33 communities across the nation. Baldwin added that the VA has been sensitive to listening to women veterans, many of whom have expressed their frustration that much of the supportive housing that is available has only one bedroom even though many veterans—women and men—are parents who need space for their children.

Uchendu then commented on the need to have a common language in the health equity field, particularly when talking about disability and vulnerable populations. She suggested that when organizations develop new interventions, they proactively build evaluation into their program in addition to setting SMART (specific, measurable, actionable, replicable, and time-bound) goals. She said that she was concerned that during the workshop she did not hear much discussion about data being broken down by race and ethnicity. At the VA, collecting data on race, ethnicity, gender, geography, age, and other demographic features is routine, which, she said, "makes it possible for various stakeholders to take their piece of the puzzle and be able to dissect it further in order to engender action." Those data also underscore the intersectionality that was mentioned repeatedly throughout the day because humans do not exist as monoliths, and they do not thrive in silos. Uchendu added that there is

---

[8] For more information, see https://www.va.gov/HEALTHEQUITY/docs/Period_of_Service_Timeline_OHE10212016.pdf (accessed March 15, 2017).

an opportunity for health care and public health to come together around the issue of housing in particular and health disparities in general. While that may seem to be a cliché, she said, in fact health care and public health are still largely operating in parallel lanes.

Uchendu also noted the importance of incorporating social determinants of health—e.g., housing, education, and economic stability—into electronic health records. While she said that she is not expecting health care providers to conduct home assessments, she said that there should be ways of outsourcing that type of information gathering. Community health workers, for example, could collect such information, as could family members. Technology could be leveraged to do this effectively, without encumbering clinical staff or impeding timely access to health care, Uchendu concluded.

DISCUSSION AND CLOSING REMARKS

Ben Bolton from the Social Security Administration asked if anyone could comment on the intersection between housing and transportation, particularly for people in non-urban areas. Baldwin said that the U.S. Department of Transportation has been doing work on this issue, as has the Association of Programs for Rural Independent Living.[9] The biggest problem, she said, is the pathway of travel. "In rural areas, you rarely have a sidewalk, and people are using their wheelchairs and walkers in the middle of the street or dirt road," she said. Street lighting is also an issue in rural communities. Community-based ride sharing, where faith-based and other local organizations make their vans accessible and provide an Uber-like service, offers one solution. With regard to health, Baldwin said that telehealth is making inroads in rural areas, and in some places doctors and practical nurses are starting to make house calls.

Carol Star from HUD said that financing the new production of affordable housing that will meet the needs of older adults and people with disabilities will remain a challenge. She noted that HUD has requested a large increase in rental assistance funds, but even if the agency is fortunate enough to get all of the funds it requested, those dollars will still be chasing a diminishing supply of housing that is accessible. In that regard, she challenged the workshop participants to think creatively about what makes sense to propose in the new environment for financing the kind of housing discussed during the workshop. Parekh suggested the possibility of combining funds from the Section 202 Supportive Housing for the

---

[9] For more information, see https://www.april-rural.org (accessed March 15, 2017).

Elderly program[10] with other funding streams, similar to the way the Section 811 Project Rental Assistance Program has done. Stone said that LeadingAge has a few providers who are mixing market rate and subsidized housing. Developers have shown an interest in this idea, though most of their efforts in many cities have been in building units for wealthier clientele. "But I do think the potential is there in terms of the sheer numbers of elderly who are going to be coming through the pipeline over the next 25 years," she said. The challenge, she added, is to identify a model that combines market rate units with more modest units and uses tax incentives to move it forward. "This might happen more at the state or local level than at the federal level," she said. "I think we are going to have to look more at private sector models with public sector investment." Lisa Sloane added that her organization has a report on how state housing agencies in Illinois, North Carolina, and Pennsylvania have pioneered new approaches to funding expansions of permanent supportive housing for extremely low-income households (O'Hara and Yates, 2015). These financing strategies, she noted, could be adapted to use funding from the Housing Trust Fund[11] program, which is targeted to this population.

Daniela Koci asked the panelists if they had seen any instances where regulators have shut down group homes that were not in full compliance with the new CMS rule regarding the settings for home- and community-based services. She also asked if they were seeing growth in different models that are more successful and more compliant. This new rule, Rosenoff said, describes the characteristics of the places that can provide Medicaid home- and community-based services, and it stresses that beneficiaries must have a choice of settings, not that they have to receive such services in their own home or in their own bedroom. She said that states are still submitting their plans to CMS with details about how they intend to comply with the rule.

A related issue, Rosenoff said, is what the U.S. Department of Justice and the states have been doing to make sure that states are complying with the Supreme Court's *Olmstead* decision. "Clearly, states have been making some progress in complying," she said, "and there have been large institutions that have shut down, some under court consent decrees, some voluntarily." In many instances, the individuals in those institutions have been moved into supportive housing and other housing models in the community. Rosenoff characterized this effort as one that is still evolving, and

---

[10] For more information, see https://portal.hud.gov/hudportal/HUD?src=/program_offices/housing/mfh/progdesc/eld202 (accessed March 15, 2017).

[11] The Housing Trust Fund is an affordable housing program at HUD. Funding under this program can be used for construction or preservation of affordable housing. For more information, see https://www.hudexchange.info/programs/htf (accessed March 15, 2017).

she said that a state will be complying if it is providing person-centered care and has a range of options for Medicaid beneficiaries and the places that can serve them. Baldwin added that the number one goal is not to close institutions but to ensure that those who live in them are not abused or neglected. She added that the concern she has heard about most from people who have moved out of institutions and into the community is that of loneliness once they leave the social environment of an institution.

Teresa Lee, providing the final comments of the day, said that she thinks that the new administration's interest in reducing regulations may provide an opportunity to break down the silos between health care and housing, both of which are heavily regulated. "To me, as I am thinking about it, we have our work cut out for us regardless of administration," Lee said. "We have our work cut out for us from a budget standpoint, and it is more important than ever to try to come together, to reframe, and to try to find common language. I encourage all of you to continue to seek out those partnerships and to engage with one another."

References

Aidala, A. A., G. Lee, D. M. Abramson, P. Messeri, and A. Siegler. 2007. Housing need, housing assistance, and connection to HIV medical care. *AIDS and Behavior* 11(6 Suppl):101-115.

Alecxih, L., S. Shen, I. Chan, D. Taylor, and J. Drabek. 2010. *Individuals living in the community with chronic conditions and functional limitations: A closer look.* Washington, DC: Office of the Assistant Secretary for Planning & Evaluation.

Baker, K., P. Baldwin, K. Donahue, A. Flynn, C. Herbert, E. La Jeunesse, M. Lancaster, I. Lew, E. Marya, K. Manning, D. McCue, J. Molinsky, R. Sanchez-Moyano, A. von Hoffman, and A. Will. 2014. *Housing America's older adults—Meeting the needs of an aging population.* Cambridge, MA: Joint Center for Housing Studies of Harvard University.

Bryant-Stephens, T., and Y. Li. 2008. Outcomes of a home-based environmental remediation for urban children with asthma. *Journal of the National Medical Association* 100(3):306-316.

Doran, G. T. 1981. There's a S.M.A.R.T. way to write management's goals and objectives. *Management Review* 70(11):35-36.

Fraze, T., V. A. Lewis, H. P. Rodriguez, and E. S. Fisher. 2016. Housing, transportation, and food: How ACOs seek to improve population health by addressing nonmedical needs of patients. *Health Affairs* 35(11):2109-2115.

Gawande, A. 2016. Health of the nation. *The New Yorker*. November 21.

Gitlin, L. N., L. Winter, M. P. Dennis, M. Corcoran, S. Schinfeld, and W. W. Hauck. 2006. A randomized trial of a multicomponent home intervention to reduce functional difficulties in older adults. *Journal of the American Geriatrics Society* 54(5):809-816.

Hoffman, D. W., and G. A. Livermore. 2012. The house next door: A comparison of residences by disability status using new measures in the American Housing Survey. *Cityscape: A Journal of Policy Development and Research* 14(1):5-33.

Houtenville, A. J., D. L. Burucker, and E. A. Lauer. 2015. *Annual compendium of disability statistics: 2015.* Durham, NH: Institute on Disability, University of New Hampshire.

Johnson, R. W. 2015. *Housing costs and financial challenges for low-income older adults.* Washington, DC: Urban Institute.

Joint Center for Housing Studies. 2016. *Projections and implications for housing a growing population: Older households 2015-2035*. Cambridge, MA: Harvard University.

Jutkowitz, E., L. N. Gitlin, L. T. Pizzi, E. Lee, and M. P. Dennis. 2012. Cost effectiveness of a home-based intervention that helps functionally vulnerable older adults age in place at home. *Journal of Aging Research* 2012:680265.

Kemper, P., H. L. Komisar, and L. Alecxih. 2005. Long-term care over an uncertain future: What can current retirees expect? *Inquiry* 42(4):335-350.

Miller, S. C., V. Mor, and J. F. Burgess, Jr. 2016. Studying nursing home innovation: The green house model of nursing home care. *Health Services Research* 51(Suppl 1):335-343.

Molinsky, J. 2016. Older households 2015-2035: Projections and implications for housing a growing population. Talk presented at Affordable and Accessible Housing for Vulnerable Older Adults and People with Disabilities Living in the Community: A Workshop, Washington, DC. http://nationalacademies.org/hmd/~/media/Files/Activity%20Files/Aging/AgingForum/2016-DEC-12/Molinsky.pdf (accessed March 16, 2017).

NIA (National Institute on Aging). 2013. *Falls and older adults*. https://nihseniorhealth.gov/falls/homesafety/01.html (accessed February 27, 2017).

O'Hara, A., and J. Yates. 2015. *Creating new integrated permanent supportive housing opportunities for ELI households: A vision for the future of the National Housing Trust Fund*. Boston, MA: Technical Assistance Collaborative.

Ruiz, S., L. P. Snyder, C. Rotondo, C. Cross-Barnet, E. M. Colligan, and K. Giuriceo. 2017. Innovative home visit models associated with reductions in costs, hosptializations, and emergency department use. *Health Affairs* 36(3):425-432.

Sandel, M., K. Phelan, R. Wright, H. P. Hynes, and B. P. Lanphear. 2004. The effects of housing interventions on child health. *Pediatric Annals* 33(7):474-481.

Sandel, M., A. Baeder, A. Bradman, J. Hughes, C. Mitchell, R. Shaughnessy, T. K. Takaro, and D. E. Jacobs. 2010. Housing interventions and control of health-related chemical agents: A review of the evidence. *Journal of Public Health Management and Practice* 16(5 Suppl):S24-S33.

Schwarcz, S. K., L. C. Hsu, E. Vittinghoff, A. Vu, J. D. Bamberger, and M. H. Katz. 2009. Impact of housing on the survival of persons with AIDS. *BMC Public Health* 9:220.

Senior Health and Housing Task Force. 2016. *Healthy aging begins at home*. Washington, DC: Bipartisan Policy Center.

Sharkey, S. S., S. Hudak, S. D. Horn, B. James, and J. Howes. 2011. Frontline caregiver daily practices: A comparison study of traditional nursing homes and the green house project sites. *Journal of the American Geriatrics Society* 59(1):126-131.

She, P., and G. A. Livermore. 2007. Material hardship, poverty, and disability among working-age adults. *Social Science Quarterly* 88(4):970-989.

Szanton, S. L. 2016. Making affordable housing accessible: Community Aging in Place—Advancing Better Living for Elders (CAPABLE) study. Talk presented at Affordable and Accessible Housing for Vulnerable Older Adults and People with Disabilities Living in the Community: A Workshop, Washington, DC. http://nationalacademies.org/hmd/~/media/Files/Activity%20Files/Aging/AgingForum/2016-DEC-12/Szanton.pdf (accessed March 16, 2017).

Szanton, S. L., R. J. Thorpe, C. Boyd, E. K. Tanner, B. Leff, E. Agree, Q. L. Xue, J. K. Allen, C. L. Seplaki, C. O. Weiss, J. M. Guralnik, and L. N. Gitlin. 2011. Community Aging in Place—Advancing Better Living for Elders: A bio-behavioral-environmental intervention to improve function and health-related quality of life in disabled older adults. *Journal of the American Geriatrics Society* 59(12):2314-2320.

Szanton, S. L., J. Roth, M. Nkimbeng, J. Savage, and R. Klimmek. 2014a. Improving unsafe environments to support aging independence with limited resources. *Nursing Clinics of North America* 49(2):133-145.

Szanton, S. L., J. W. Wolff, B. Leff, R. J. Thorpe, E. K. Tanner, C. Boyd, Q. Xue, J. Guralnik, D. Bishai, and L. N. Gitlin. 2014b. CAPABLE trial: A randomized controlled trial of nurse, occupational therapist and handyman to reduce disability among older adults: Rationale and design. *Contemporary Clinical Trials* 38(1):102-112.

Szanton, S. L., J. L. Wolff, B. Leff, L. Roberts, R. J. Thorpe, E. K. Tanner, C. M. Boyd, Q. L. Xue, J. Guralnik, D. Bishai, and L. N. Gitlin. 2015. Preliminary data from Community Aging in Place—Advancing Better Living for Elders, a patient-directed, team-based intervention to improve physical function and decrease nursing home utilization: The first 100 individuals to complete a Centers for Medicare & Medicaid Services innovation project. *Journal of the American Geriatrics Society* 63(2):371-374.

Szanton, S. L., B. Leff, J. L. Wolff, L. Roberts, and L. N. Gitlin. 2016. Home-based care program reduces disability and promotes aging in place. *Health Affairs* 35(9):1558-1563.

Appendix A

Workshop Agenda

AFFORDABLE AND ACCESSIBLE HOUSING FOR VULNERABLE OLDER
ADULTS AND PEOPLE WITH DISABILITIES LIVING IN THE COMMUNITY

PUBLIC WORKSHOP

FORUM ON AGING, DISABILITY, AND INDEPENDENCE
AND
ROUNDTABLE ON THE PROMOTION OF HEALTH EQUITY AND THE ELIMINATION OF
HEALTH DISPARITIES

National Academy of Sciences Building
2101 Constitution Avenue, NW, Room 120
Washington, DC 20418

December 12, 2016

Workshop Objectives:

- Summarize the state of the knowledge on housing as a social determinant of health and as a platform for health and independence for vulnerable older adults and individuals with disabilities.
- Highlight successful and promising collaborations to provide affordable and accessible housing, with consideration for differences between rural, suburban, and urban settings.
- Explore sustainable and scalable strategies, policies, and practices to support linking affordable housing with services to benefit health and optimize independence.
- Identify data needs and research gaps to measure effectiveness of models of housing with supportive services for vulnerable older adults and individuals with disabilities.

Public Workshop: Room 120

8:30–8:35 a.m. **Welcome and Opening Remarks**

> TERESA LEE
> WORKSHOP PLANNING COMMITTEE CHAIR
> ALLIANCE FOR HOME HEALTH QUALITY AND INNOVATION

8:35–9:45 a.m. **Keynote Panel**

> *Facilitator*
> TERESA LEE
> WORKSHOP PLANNING COMMITTEE CHAIR
> ALLIANCE FOR HOME HEALTH QUALITY AND INNOVATION
>
> *Keynote Speakers (20 minutes each)*
>
> LISA MARSH RYERSON
> AARP FOUNDATION
>
> ERIKA POETHIG
> URBAN INSTITUTE
>
> *Keynote Speakers Q&A with Audience (20 minutes)*

9:45–10:55 a.m. **Panel 1: Affordability of and Financing for Housing That Supports Health and Independence for Vulnerable Older Adults and People with Disabilities**

> *Facilitator*
> GLEN WHITE
> WORKSHOP PLANNING COMMITTEE MEMBER
> UNIVERSITY OF KANSAS
>
> *Panel 1 Speakers (20 minutes each)*
>
> PURVI SEVAK
> MATHEMATICA POLICY RESEARCH
>
> JEN MOLINSKY
> HARVARD UNIVERSITY

APPENDIX A

Panel 1 Speakers Q&A with Audience (25 minutes)

10:55–11:10 a.m. **BREAK**

11:10 a.m–
12:25 p.m.
Panel 2: Design Features of Accessible Housing for Older Adults and People with Disabilities

Facilitator
ELENA FAZIO
WORKSHOP PLANNING COMMITTEE MEMBER
ADMINISTRATION FOR COMMUNITY LIVING

Panel 2 Speakers (15 minutes each)

BRYCE WARD
UNIVERSITY OF MONTANA

CORNEIL MONTGOMERY
HABITAT FOR HUMANITY INTERNATIONAL

PATRICIA TEDESCO
VERMONT CENTER FOR INDEPENDENT LIVING

Panel 2 Speakers Q&A with Audience (25 minutes)

12:25–1:25 p.m. **LUNCH**

1:25–3:25 p.m. **Panel 3: Promising Models Connecting Affordable Housing and Services as a Platform for Health and Independence**

Facilitator
CRAIG RAVESLOOT
WORKSHOP PLANNING COMMITTEE MEMBER
UNIVERSITY OF MONTANA

Panel 3 Speakers (12 minutes each)

PEGGY BAILEY
CENTER ON BUDGET AND POLICY PRIORITIES

KATINA WASHINGTON
U.S. DEPARTMENT OF HOUSING AND URBAN DEVELOPMENT

LISA SLOANE
TECHNICAL ASSISTANCE COLLABORATIVE

SARAH SZANTON
JOHNS HOPKINS SCHOOL OF NURSING

MOLLY DUGAN
CATHEDRAL SQUARE

ROSE ENGLERT
CareOregon

Panel 3 Speakers Moderated Discussion on What Makes an Affordable Housing Program a Successful Platform for Health and Independence (30 minutes)

Panel 3 Speakers Q&A with Audience (25 minutes)

3:25–3:40 p.m.	**BREAK**
3:40–4:30 p.m.	**Reactors Panel on Policy Implications and Research Needs**

Facilitator
EMILY ROSENOFF
WORKSHOP PLANNING COMMITTEE MEMBER
OFFICE OF THE ASSISTANT SECRETARY FOR PLANNING AND EVALUATION
U.S. DEPARTMENT OF HEALTH AND HUMAN SERVICES

Reactors
DARA BALDWIN
NATIONAL DISABILITY RIGHTS NETWORK

ANAND PAREKH
WORKSHOP PLANNING COMMITTEE MEMBER
BIPARTISAN POLICY CENTER

                    Robyn Stone
                    LeadingAge

                    Uchenna S. Uchendu
                    Workshop Planning Committee Member
                    U.S. Department of Veterans Affairs

                    Reactor Panel Speakers Q&A with Audience (20 minutes)

4:30–4:45 p.m.   **Closing Remarks**

                    Teresa Lee
                    Workshop Planning Committee Chair
                    Alliance for Home Health Quality and Innovation

4:45 p.m.   **Adjourn**

Appendix B

Biographical Sketches of Workshop Speakers and Reactors

Peggy Bailey, M.P.A., is the director of the Health Integration Project at the Center on Budget and Policy Priorities, where she identifies opportunities to improve health care policy to better link with housing programs, serves those involved in the criminal justice system, improves quality and access to behavioral health services, and incorporates human services needed by vulnerable populations. Ms. Bailey's career includes work on federal, state, and local policy and service delivery on a wide variety of issue areas, including Medicaid eligibility and benefits for families and people with disabilities, public health innovation, behavioral health service delivery and integration with primary care, youth homelessness policy and service delivery, and child welfare. Prior to joining the center in January 2016, she was the director of health systems integration for the Corporation for Supportive Housing. She has also worked for the National Alliance to End Homelessness, the Association of Maternal and Child Health Programs, and the City of Rockwall, Texas. Ms. Bailey holds a bachelor of arts degree from the University of Notre Dame and a master of public affairs degree from The University of Texas at Dallas.

Dara Baldwin, M.P.A., is the senior public policy analyst for the National Disability Rights Network (NDRN) in Washington, DC. NDRN, the nonprofit membership organization for the federally mandated Protection and Advocacy Systems and Client Assistance Programs, is the largest provider of legally based advocacy services to people with disabilities in the United States. She works on NDRN's diversity and cultural competency team and is responsible for outreach as well as working on coalitions to assist with better legislative outcomes for the community. She has extensive knowledge of the Americans with Disabilities Act of 1990 (ADA) and other disability laws. She has a keen ability for networking and outreach to grassroots, national, and international advocates. She has led multiple national and international advocacy campaigns. Prior to this position, Ms. Baldwin was an ADA compliance specialist in the DC government, a policy analyst at the National Council on Independent Living, a child advocate in New Jersey, and a senior policy analyst on criminal justice issues. She serves on the board of directors for the National Low Income Housing Coalition and has served as a trustee on the American Society for Public Administration's board of insurance trustees for two terms. She has a bachelor of arts in political science from Rutgers University and was a Pi Alpha Alpha honors graduate with a master's of public administration from Rutgers University School of Public Affairs and Administration.

Molly Dugan, M.P.A., started at Cathedral Square in October 2008, focusing primarily on new housing development, but also assisting with development of the Support And Services at Home (SASH) initiative. When the program design of SASH started full bore in July 2009, Ms. Dugan became statewide SASH director. Prior to Cathedral Square, she worked for the state of Vermont in the Community Development Program at the Department of Housing and Community Affairs (DHCA). She served as a senior community development specialist and then the director. Ms. Dugan became deputy commissioner of DHCA in August 2006 and then served as DHCA's acting commissioner. She received a B.S. in economics and political science from the University of New Hampshire and an M.P.A. from the University of Vermont.

Rose Englert leads CareOregon's Community Health Innovation Program department. The team aligns the clinical care of low-income Medicare and Medicaid members with tailored supports and services that address social determinants of health. With a focus on housing, food and nutrition, social support, and transportation, these interventions are showing positive early outcomes and defining the role of health plans within the social services realm. A health care professional with 17 years of public health, policy, and contracting experience, Ms. Englert has been active in the operations of a broad range of health-related nonprofits, corporations, associations, and professional alliances in multiple states.

Jennifer Molinsky, Ph.D., M.P.A., is a senior research associate at the Joint Center for Housing Studies of Harvard University, where she manages the Joint Center's work on housing for older adults. She was lead author on *Older Households 2015–2035: Projections and Implications for Housing a Growing Population* (2016) as well as *Housing America's Older Adults: Meeting the Needs of an Aging Population* (2014). Dr. Molinsky's work also touches on land use regulation, multifamily housing, and family-sized housing supply. She was a co-editor of the 2014 book *Homeownership Built to Last: Balancing Access, Affordability, and Risk After the Housing Crisis*. She is also a lecturer at Harvard's Graduate School of Design. Prior to joining the Joint Center, Dr. Molinsky served as the chief planner for long-range planning in Newton, Massachusetts; a researcher at the Lincoln Institute of Land Policy; the associate director of issues at the Municipal Art Society of New York; and as a member of the planning board in Cambridge as well as other local planning committees. She has also held positions with Abt Associates and with PricewaterhouseCoopers' government housing finance practice, where she worked on projects related to housing finance, affordable housing, and community development. She holds a Ph.D. in urban planning from the Massachusetts Institute of Technology, a master's of public affairs, urban and regional planning, from the Woodrow Wilson School at Princeton University, and a B.A. from Yale University.

Corneil Montgomery, Ph.D., is a senior program specialist at Habitat for Humanity International, where he is plays a key role in the design and implementation of the neighborhood revitalization program and strategy. In addition, he provides leadership and serves as principal strategist for Habitat's aging in place strategy. With more than 7 years of experience in nonprofit and community development, Dr. Montgomery has expertise in combating complex and interrelated issues such as aging in place, housing, community stabilization, education, youth development, and more. In 2016, Dr. Montgomery obtained his doctoral degree in public policy and administration with a specialization in local government management for sustainable communities from Walden University. His doctoral research was titled "Adopting the Lifelong Communities Initiative in the Atlanta Metropolitan Area," which entailed exploring the processes involved in planning and implementing an aging in place initiative at the regional and local levels. He is passionate about and committed to advocacy for and the development of sustainable and livable communities for people of all ages and abilities.

Anand Parekh, M.D., M.P.H., is the Bipartisan Policy Center's (BPC's) chief medical advisor, providing clinical and public health expertise across the organization, particularly in the areas of aging, prevention, and global health. Prior to joining BPC, he completed a decade of service at the U.S. Department of Health and Human Services (HHS). As the deputy assistant secretary for health from 2008 to 2015, he developed and implemented national initiatives focused on prevention, wellness, and care management. Briefly in 2007, he was delegated the authorities of the assistant secretary for health overseeing 10 health program offices and the U.S. Public Health Service Commissioned Corps. Earlier in his HHS career, he played key roles in public health emergency preparedness efforts as special assistant to the science advisor to the secretary. Dr. Parekh is a board-certified internal medicine physician, a fellow of the American College of Physicians, and an adjunct assistant professor of medicine at Johns Hopkins University, where he previously completed his residency training in the Osler Medical Program of the Department of Medicine. He provided volunteer clinical services for many years at the Holy Cross Hospital Health Center, a clinic for the uninsured in Silver Spring, Maryland. Dr. Parekh is an adjunct professor of health management and policy at the University of Michigan School of Public Health. He currently serves on the dean's advisory board of the University of Michigan School of Public Health, the Presidential Scholars Foundation board of directors, and the board of directors of WaterAid America. He has spoken widely and written extensively on a variety of health topics such as chronic care management, population health, value in health care, and the need for health and human services integration. A native of Michigan, Dr. Parekh received a B.A. in political science, an M.D., and an M.P.H. in health management and policy from the University of Michigan. He was selected as a U.S. Presidential Scholar in 1994.

Erika Poethig, M.A., is an institute fellow and the director of urban policy initiatives at the Urban Institute. She leads the policy advisory group, which assembles Urban Institute experts to help leaders draw insights from research and navigate policy challenges facing urban America. She also leads partnerships to develop new programs and strategies, translate research into policy and practice, and align philanthropic investments and federal policy. Before joining Urban, Ms. Poethig was the acting assistant secretary for policy, development, and research at the U.S. Department of Housing and Urban Development. During her tenure in the Obama administration, she was also the deputy assistant secretary for policy development and was a leading architect of the White House Council for Strong Cities and Strong Communities. At the John D. and Catherine T. MacArthur Foundation, she was the associate director for housing. She also was the assistant commissioner for policy, resource, and program development at the City of Chicago's Department of Housing. In the late 1990s, she developed Mayor Richard Daley's campaign to combat predatory lending, prevent foreclosures, and stabilize communities. Previously, she was an associate project director of the Metropolis Project, which produced the Metropolis 2020 agenda for regional leadership around the major issues faced by the metropolitan Chicago area. Ms. Poethig serves on the board of the Center for Community Progress and the Wooster board of trustees. Ms. Poethig was a Phi Beta Kappa member at the College of Wooster, a Fulbright Scholar at the University of Vienna, and has an M.A. with honors in public policy from the University of Chicago.

Emily Rosenoff, M.P.A., is with the U.S. Department of Health and Human Services Office of the Assistant Secretary for Planning and Evaluation in the Office of Disability, Aging and Long-Term Care Policy. Her work includes a focus on homelessness, housing with services, residential care and assisted living policy, and Medicaid home and community-based services policy. Ms. Rosenoff has also worked at the Organisation for Economic Co-operation and Development, the Senate Committee on Veterans' Affairs, and the University of California, San Francisco's Center for the Health Professions. She received a master's in public affairs from Princeton University's Woodrow Wilson School and her bachelor's degree in molecular and cell biology from the University of California, Berkeley.

Lisa Ryerson, M.S., has served as the president of AARP Foundation, AARP's affiliated charity, since 2013. She is an experienced and innovative leader who is responsible for setting the foundation's strategic direction and for leading the organization's efforts to create opportunities for older Americans struggling with poverty and the related issues of hunger, income, housing, and social isolation. Ms. Ryerson has received numerous awards and honors for her leadership and service, both at the foundation and in previous positions. She has elevated the foundation's visibility through innovative collaborations with other organizations, such as the National Football League's Miami Dolphins. Under her leadership, the foundation has secured unprecedented funding to help provide programs and services that truly change lives. Before joining AARP Foundation, Ms. Ryerson served as the president and chief executive officer of Wells College in Aurora, New York.

Purvi Sevak, Ph.D., is a professor in the Economics Department at Hunter College of the City University of New York and a senior researcher at Mathematica Policy Research. Her research focuses on public policy and labor market outcomes for vulnerable populations including individuals with disabilities, older workers, and immigrants. Her work has been supported by the Social Security Administration, the U.S. Department of Education (National Institute on Disability and Rehabilitation Research), and the U.S. Department of Health and Human Services (National Institute on Disability, Independent Living, and Rehabilitation Research). Professor Sevak holds a Ph.D. in economics from the University of Michigan. She is a member of the Association for Public Policy Analysis and Management and the American Economic Association.

Lisa Sloane, M.P.A., has more than 25 years of experience working with federal, state, and local governments as well as nonprofit agencies to address the supportive housing needs of people with disabilities, including homeless individuals and their families. Her work in the area of ending homelessness includes the development and implementation of training as well as technical assistance to continuums of care and individual programs. She has also worked with the states of Pennsylvania and Louisiana to develop and implement permanent supportive housing programs. In Massachusetts, she played a key role in the development of innovative cross-disability housing programs, including a housing locator system, a state housing bond fund, and a state home modification loan program. She has expertise in the area of fair housing. Prior to joining Technical Assistance Collaborative (TAC), Ms. Sloane was principal of Sloane Associates, a woman-owned business that provided affordable housing and human services consultation, specializing in the development of housing programs and policies for persons with disabilities, including homeless persons with disabilities.

Robyn I. Stone, Ph.D., a noted researcher and internationally recognized authority on long-term care and aging policy, is the senior vice president for research at LeadingAge and the executive director of the LeadingAge Center for Applied Research. She has held senior research and policy positions in both the U.S. government and the private sector. She was a political appointee in the Clinton administration, serving in the U.S. Department of Health and Human Services as the deputy assistant secretary for disability, aging and long-term care policy and the assistant secretary for aging. Dr. Stone is a distinguished speaker and has been published widely in the areas of long-term care policy and quality, chronic care for people with disabilities, aging services workforce development, low-income senior housing, and family caregiving. She serves on numerous provider and nonprofit boards that focus on aging issues. Dr. Stone is a fellow of the Gerontological Society of America and the National Academy of Social Insurance. She was elected to the National Academy of Medicine in 2014.

Sarah L. Szanton, Ph.D., M.S.N., is an associate professor at the Johns Hopkins School of Nursing with a joint appointment in the Department of Health Policy and Management at the Johns Hopkins Bloomberg School of Public Health. She tests interventions to reduce health disparities among older adults. Her work particularly focuses on ways to help older adults age in place as they grow older. These include ways to improve the social determinants of health such as modifying housing and improving access to food. Dr. Szanton completed undergraduate work in African-American Studies at Harvard University and earned a bachelor's degree from the Johns Hopkins School of Nursing in 1993. She holds a nurse practitioner master's degree from the University of Maryland and a doctorate from Johns Hopkins University. She is the associate director for policy at the Center for Innovative Care in Aging at Johns Hopkins as well as core faculty at the Center on Aging and Health at the Hopkins Center for Health Disparities Solutions, adjunct faculty with the Hopkins Center for Injury Research and Policy, and adjunct faculty at Arizona State University. She has been by funded by AARP Foundation, the Centers for Medicare & Medicaid Services Innovation Center, the Hillman Foundation, The John A. Hartford Foundation, the National Institutes of Health, and the Robert Wood Johnson Foundation.

Patricia Tedesco is the coordinator of the Home Access Program at the Vermont Center for Independent Living (VCIL). VCIL, established in 1979, is a statewide nonprofit organization directed and staffed by individuals with disabilities which works to promote the dignity, independence, and civil rights of Vermonters with disabilities. The Home Access Program has been in place for 33 years. Ms. Tedesco joined the VCIL staff 5 years ago and has been the Home Access Program coordinator since 2014. Responsible for overseeing access modifications for Vermonters with disabilities, Ms. Tedesco works with access consultants and independent contractors statewide who receive training in the Americans with Disabilities Act specifications. She has established relationships with other nonprofit organizations, Medicaid programs, and the five NeighborWorks organizations in Vermont to secure leverage funding. The modifications, primarily installing ramps and accessible bathrooms, support Vermonters to stay in their homes and communities and thus stay out of institutional settings. By aging in place, members of this vulnerable population, often living in rural areas of the state, have better access to social opportunities and medical care and can therefore live a more engaged life. Ms. Tedesco is a member of the Vermont Affordable Housing Coalition and the National Council on Independent Living. Prior to working in the field of human services and advocacy, Ms. Tedesco successfully operated a photography business for 23 years. In addition to numerous awards for her photography, Ms. Tedesco was awarded the Young Careerist Award by the State of Vermont's Business and Professional Women's organization. She earned a business degree from Endicott College and a degree in liberal studies from the State University of New York at Brockport.

Uchenna S. Uchendu, M.D., was appointed as the chief officer for health equity at the Veterans Health Administration (VHA) within the U.S. Department of Veterans Affairs in 2013. In this role, Dr. Uchendu is championing the advancement of health equity and the reduction of health disparities for all, especially vulnerable populations based on racial or ethnic group; religion; socioeconomic status; gender; age; mental health; cognitive, sensory, or physical disability; sexual orientation; geographic location; military era or other characteristics historically linked to discrimination or exclusion. Dr. Uchendu launched and steered the VHA Health Equity Coalition to deliver the first VHA Health Equity Action Plan in record time and is leading the charge for the deployment. She represents VHA and serves as liaison to other governmental and nongovernmental organizations working to achieve health equity. Dr. Uchendu is a general internist who has practiced in multiple settings. She is a member of the American College of Physicians and the American Association for Physician Leaders.

Bryce Ward, Ph.D., is a associate director and the director of health care research at the Bureau of Business and Economic Research at the University of Montana. Dr. Ward holds a Ph.D. in economics from Harvard University and a bachelor's degree in economics and history from the University of Oregon. He works across a wide variety of topics within the broad areas of econometric analysis and applied microeconomics, including urban and regional economics, labor economics, health economics, public finance, and environmental and natural resource economics. Recently, with collaborators at the University of Montana's Rural Institute for Inclusive Communities, Dr. Ward has focused on understanding the relationships between housing, the environment, health, and participation for people with disabilities.

Katina Washington is a program analyst in the Grants and New Funding Branch within the U.S. Department of Housing and Urban Development (HUD), Office of Multifamily, Office of Asset Management and Portfolio Oversight, Assisted Housing Oversight Division. In this position, she serves as the program lead for the Section 811 Project Rental Assistance (811 PRA) program and manages the implementation of the 811 PRA program. Ms. Washington has more than 8 years of federal grants management experience previously serving as the lead for the Assisted Living Conversion Program. She has also has experience with working on policy matters within multifamily housing assisted programs. Prior to HUD, Ms. Washington worked in program administration and adolescent health care programs for Prince Georges County in Maryland. Ms. Washington received a B.S. in business administration from Bowie State University.